CLIMATE CHANGE AND NEW CHALLENGES

ABOUT THE EDITORS

Prof. Dr. Vir Singh holds dual Ph.D.: Animal Nutrition (Fodder Resources) from G.B. Pant University, and Botany (Rangeland Management) from H.N.B. Garhwal University. Earlier he also completed Master's degrees in Animal Nutrition, Botany, and Sociology, and a Diploma in Environmental Management from Israel. Currently he is a Professor at the Department of Environmental Science, G.B. Pant University of Agriculture and Technology. He has wide experience of teaching, research, extension, planning and journalism. He has travelled many parts of South Asia, Middle East, and Europe. He has authored and edited 22 books and published more than 200 articles.

Dr. Govind Singh Kushwaha is presently an Assistant Professor of Psychology, Department of Social Sciences and Humanities, G.B. Pant University of Agriculture and Technology, Pantnagar. He has attended a large number of conferences and presented erudite research papers. His area of interest is Environmental Psychology, Stress, Clinical and Social Psychology.

Bipin Kumar is a Research Scholar at the Department of Environmental Science, C.B.S.H., G.B. Pant University of Agriculture and Technology, Pantnagar. He carried out a prestigious project on community based ecotourism. Currently, he is engaged in a project on bioremediation of heavy metals from industrial effluents using indigenous microorganisms.

CLIMATE CHANGE AND NEW CHALLENGES

Society, Environment and Development

Edited by

Vir Singh
Govind Singh Kushwaha
Bipin Kumar

CONCEPT PUBLISHING COMPANY PVT. LTD., NEW DELHI-110059

ISBN-13: 978-81-8069-866-8

First Published 2012

Published and Printed by

Concept
Since 1974

Concept Publishing Company Pvt. Ltd.
A/15-16, Commercial Block, Mohan Garden
New Delhi-110059 (India)
T: 25351460, 25351794, *F*: 091-11-25357109
E: publishing@conceptpub.com, W: www.conceptpub.com
Editorial Office: H-13, Bali Nagar, New Delhi-110 015, India.

Cataloging in Publication Data--Courtesy: D.K. Agencies (P) Ltd. <docinfo@dkagencies.com>

Climate change and new challenges : society, environment and development edited by Vir Singh, Govind Singh Kushwaha, Bipin Kumar.
p. cm.
Includes bibliographical references and index.
ISBN 9788180698668

1. Climatic changes—Social aspects—India. 2. Human beings—Effect of climate on—India. 3. Climatic changes—Health aspects—India. 4. Community development—India. 5. Sustainable development—India. 6. Community forestry—India. I. Singh, Vir. II. Kushwaha, Govind Singh. III. Bipin Kumar.

DDC 363.738740954 23

Preface

Climate change is undoubtedly the hottest issue of our times. It also remains to be the most formidable challenge facing humanity, for an unfavourable change in the climate system of the biosphere beyond a certain point would not only pose a threat to the sustainability of our production systems but also to the very existence of life on Earth. There could be several reasons of global warming, but anthropogenic reason appears to be the dominant one. Yes, planet Earth is in unprecedented crises owing to human activities. What a paradox! In order to exist, sustain, progress, be happy, and do everything, humans cannot do without a living and vibrant biosphere. Humans are not only a part, but a flowering of the evolution of life on Earth.

Climate order of the Earth too is a state of the evolution creating conditions responsible for the phenomenon of life and for maintaining the planet life-conditioned. Breaking down the processes of life-conditioning begins with putting the climate order of biosphere into instability and chaos. Once the biosphere is in the state of climate instability and climate chaos, it ceases to be life-ameliorating. It rather turns life-depressing. And in a long run, it ought to turn life-annihilating.

Life-embracing, life-enhancing, life-supporting and life-sustaining processes going on in an ecosystem are linked with our socio-economic progress, social happiness and with our own sustainability. Life-dispelling, life-eclipsing, life-depressing and life-teasing processes, on the other hand, lead to socio-economic collapse. In other words, ecological affluence is socio-economic affluence; ecological coup is socio-economic coup. Ecosystem functions, thus, determine socio-economic status of a country. Well-being of biosphere is the well-being of the global society. Well-being of the biosphere and climate order of the world go hand in hand. And this is the essence of our progress, of our happiness, of our sustainability.

As the climate chaos and climate instability leading to climate change are intensifying, new challenges facing humanity are emerging and posing threats of various sorts. At the same time, as the consciousness about adverse climate change is growing the world over, new perspectives and new adaptation and mitigation strategies are coming to the fore. The book in your lotus hands is also a product of such consciousness about our environment, about our biosphere, about our Earth. We are with the majority of thinkers who think that the adverse climate change being experienced is owing to anthropogenic reasons. We also believe that with the most conscious and honest efforts, we can heal our biosphere, we can help our planet to restore its climate order.

We hope that this book will further stimulate and inspire our inner being and contribute to promote human efforts in the sacred cause of mitigating climate change. We would be grateful to readers and users of the book if they could provide their valuable suggestions leading to better care of our environment, of our planet.

Collection, selection and editing of articles as chapters in the book constituted a big project in itself. We wish to express our gratitude to all who directly or indirectly gave their input vital for bringing the book in this form. First of all, we would like to express our gratitude and appreciation to all the contributing authors, who shared their views, perspectives, and strategies with us as common commitment to Mother Earth. We are also grateful to our Vice Chancellor Dr. B.S. Bisht; Dean, College of Basic Sciences and Humanities Dr. B.R.K. Gupta; Head, Environmental Science Dr. Uma Melkania; prominent environmentalists Dr. J.P.N. Rai and R.K. Srivastava. Research Scholar Nanda Nautiyal gave her valuable time and inputs in bringing the book in this form. Our thanks are also due to Akanksha Rastogi, Anil Kumar Shankhwar, Kiran Rawat, Megha Bora, Zeba Khanam, Kavita Tariyal, Parul Sundha, Mohini Singh, Bhumija, Anand Prabha Rawat, Krishna, Vrinda Negi, Suman Lata Panwar, Monika, Deepika, Roopam, Pooja, Tripti, Amit, Megha Verma, Meenakshi and Sandhya for various kinds of inputs necessary for accomplishing this task.

We should like to thank Shri Ashok Kumar Mittal of Concept Publishing Company Pvt. Ltd., New Delhi for agreeing to publish this edited work and for valuable suggestions which find articulation in this work. Numerous titles of the Concept Publishing Company Pvt. Ltd. have shot into fame. I hope this book would prove itself yet another feather in the cap of the prominent Indian publisher.

The views and ideas expressed in different chapters are those of the authors and may not necessarily corroborate with those of the editors and publisher. Further, the editors and publisher are not responsible for the source the authors depended upon to analyse their views and/or derive conclusions.

We hope the book proves instrumental in meeting the gravest challenge of our times – formidable climate change—and in healing our planet.

Vir Singh
Govind Singh Kushwaha
Bipin Kumar

Contents

Part II
FROM THE CLIMATE OF DESPAIR TO THE CLIMATE OF HOPE
Existing Potentials to Respond Maldevelopment

Part III
CLIMATE OF CHANGE
Changing to Challenge Climate Change

List of Contributors

Ali Seid Mohammed, Department of Biology, Bahir Dar University, Ethiopia; alinabiot@yahoo.com, Ph.: +251-918767869.

Amit Kumar, Rajiv Gandhi National University of Law, The Mall, Mahendra Kothi, Patiala-147001, Punjab.

Anjali Adhikari, College of Agribusiness Management, G.B. Pant University of Agriculture and Technology, Pantnagar-263145; adhikarianjali@gmail.com

Anjali Dewan, Department of Home Science, St. Bede's College, Shimla, Himachal Pradesh; dewananjali@rediffmail.com

Aparna Rajput, Department of Agriculture Communication, College of Agriculture, G.B. Pant University of Agriculture and Technology, Pantnagar-263145; aparnarajput7@gmail.com

Arpita Sharma, Deptt. of Agricultural Communication, G.B. Pant University of Agriculture and Technology, Pantnagar-263145 sharmaarpita35@gmail.com

Ashima Srivastava, Department of Chemistry, JSS Academy of Technical Education, Noida; ashima14dec@rediffmail.com

Ashish Tewari, Uttarakhand Centre for Climate Change (UCCC), Kumaun University, Nainital-263102, Uttarakhand.

B. Singh, Extension Education, Institute of Home Science, Agra.

Bipin Kumar, Gurukula Kangri Vishwavidyalaya, Haridwar; envbipinkumar@gmail.com

Bitapi C. Sinha, Senior Scientist, Wildlife Institute of India, Dehradun.

Chitra Singh, Deptt. of Elect. and Comm. Engg, College of Technology, G.B. Pant University of Agriculture and Technology, Pantnagar-263145; chitra.singh88@gmail.com

D. Jayadeva Gowda, A.V.K. College for Women, Hassan-573201, Karnataka.

Deepali Joshi, College of Agribusiness Management, G.B. Pant University of Agriculture and Technology, Pantnagar-263145 joshideepali@gmail.com

Deepika Singh, Career Institute of Technology and Management, Faridabad-121 001.

Dinesh Bhatt, Department of Zoology, Gurukula Kangri Vishwavidyalaya, Haridwar.

Gaurav Papnai, Deptt. of Agricultural Communication, College of Agriculture, G.B. Pant University of Agriculture and Technology, Pantnagar-263145; gauravpapnai@gmail.com

Gita Singh, Columbus Public School, Rudrapur, Uttarakhand; gitavirsingh@gmail.com

Govind Singh Kushwaha, Department of Social Science and Humanities, College of Basic Sciences and Humanities, G.B. Pant University of Agriculture and Technology, Pantnagar-263145; drgovindsingh@yahoo.co.in

Hema Bhatt, Deptt. of Family Resource Management, G.B. Pant University of Agriculture and Technology, Pantnagar-263145; hemabhatt2000@gmail.com

Kavita Bhatnagar, Department of Economics, Gokuldas Girls Degree College, Moradabad-244001.

Kiran Yadav, Deptt. of Agricultural Communication, College of Agriculture, G.B. Pant University of Agriculture and Technology, Pantnagar-263145.

Kush Kalra, Rajiv Gandhi National University of Law, The Mall, Mahendra Kothi, Patiala-147001, Punjab.

Nandana Kumari, Department of Foods and Nutrition, College of Home Science, G.B. Pant University of Agriculture and Technology, Pantnagar-263145.

Neelam Bhardwaj, Deptt. of Agricultural Communication, College of Agriculture, G.B. Pant University of Agriculture and Technology, Pantnagar-263145; n_11bhard@yahoo.co.in

Nisha Singh, New Delhi Institute of Management Studies, New Delhi, 110020.

Pratibha Singh, Department of Chemistry, J.S.S. Academy of Technical Education, Noida.

R.P. Singh, Deptt. of Elect. and Comm. Engg, College of Technology, G.B. Pant University of Agriculture and Technology, Pantnagar-263145; dr.ravipratapsingh@gmail.com

Rita Patnaik, Department of Foods and Nutrition, College of Home Science, G.B. Pant University of Agriculture and Technology, Pantnagar-263145, Uttarakhand.

Rita Singh Raghuvanshi, Department of Foods and Nutrition, College of Home Science, G.B. Pant Univ. of Agriculture and Technology, Pantnagar-263145; deanhsc1@gmail.com

Roli Verma, Department of Chemistry, J.S.S. Academy of Technical Education, Noida.

Savita Rathour, Deptt. of HDFS, G.B. Pant University of Agriculture and Technology, Pantnagar-263145; savitarathour@yahoo.com

Shalini, Home Science, SRDAKPG College, Hathras, Uttar Pradesh,

Shiwani Bhatnagar, Department of Entomology, College of Agriculture, G.B. Pant University of Agriculture and Technology, Pantnagar-263145; shiwanibhatnagar@rediffmail.com

Shreena Keswani, College of Agribusiness Management, G.B. Pant University of Agriculture and Technology, Pantnagar-263145; shreenakeshwani@gmail.com

U.P.S. Bhardwaj, Deptt. of Elect. and Comm. Engg, College of Technology, G.B. Pant University of Agriculture and Technology, Pantnagar-263145; uday4u007@gmail.com

Uma Melkania, Department of Environmental Science, College of Basic Sciences and Humanities, G.B. Pant University of Agriculture and Technology, Pantnagar-263145.

Vardan Singh Rawat, Uttarakhand Centre for Climate Change (UCCC), Kumaun University, Nainital-263102, Uttarakhand.

Vir Singh, Department of Environmental Science, College of Basic Sciences and Humanities, G.B. Pant University of Agriculture and Technology, Pantnagar-263145; drvirsingh@rediffmail.com

Y.D. Mishra, KVK, Gwaldam, G.B. Pant University of Agriculture and Technology, Pantnagar-263145.

Y.S. Rawat, Uttarakhand Centre for Climate Change (UCCC), Kumaun University, Nainital-263102, Uttarakhand.

Introduction

Climate Change Chaos : Causes, Concerns and Challenges

Vir Singh, Govind Singh Kushwaha and *Bipin Kumar*

"Development" is not just a notion of economic growth which has become an imperative of human progress. It is a phenomenon of *Homo sapiens*, vital for the very existence of human race on Earth. Man can live without anything but not without development. He breathes development, he drinks development, he eats development. Development is the core truth of human life. Man is born for development. He takes care for development. But, haplessly, development does not take care for human beings. Man thinks and does everything about development. Long strides in science and advances in technology have been used to nourish development, so much so that man now no more lives in biosphere, but in a technosphere which he himself has created as an 'alternative' to biosphere.

But can technosphere be an alternative to biosphere? Yes, it can, but for how long? Only for a limited certain period of time. Life, however, is not countable in number of days, or for a certain period of time. Life is sustainable. Life is eternal. For the life to be sustainable and eternal, life needs to be accommodated in biosphere, not in technosphere. Technosphere is bound to be lifeless one day, at a point of time, earlier than a distant future.

Terminator of "Technosphere"

Technosphere is a phenomenon of development—development that lives unto itself. It does not serve human needs on a sustained basis.

Development has shallow roots and is doomed to be collapsed. Development does not collapse alone. Since it is a phenomenon, it is bound to collapse with life-support systems: forests, rivers, lakes, rangelands, cultures, and what not! Human race too is in its grab. But the point is to come when human race will be in the firm clutches of development. Signals to this effect are coming—in the form of climate change leading to warming of the planet, shrinking of natural resources, extinction of species, increasing uncertainties and human miseries, and the likes.

Climate change chaos our living Earth is currently facing a peculiar and predictable scenario which has come out of the womb of development. Development has already killed biosphere by replacing it with technosphere. And now technosphere is dying slow death. Climate change chaos is the obvious terminator of the life of technosphere. Yes, ominous changes in the climate systems should be regarded as a terminating process to lead to the collapse of the technosphere. Technosphere itself is a terminator of the living biosphere.

Reviewing the Conventional Development Frameworks

The first thing to return to the biosphere is to review the conventional development. The climate change is rooted into development paradigm. Therefore, we need to observe a shift in development paradigm. Development devoid of environmental considerations leads towards destruction. Trees by trees, forests by forests, rivers by rivers, the whole system is shrinking to a critical point and appears to be on the verge of collapse. The conventional development frameworks need sacrifice of ecological systems. In that way economic growth is maintained. Overexploited for increasing 'growth', the ecosystems stay in the state of unsustainability.

The sustainable development concept calling for the vital integration of development and environment has now come of age. But the concept evolved in the Brundtland Report—*Our Common Future*—in 1987 hardly finds articulation. Destruction-led development, on the other hand, is going on at a faster pace than ever in the past.

"It is only through new sustainable approaches to development that the planet's fragile ecosystems can be protected and the aims of human development be furthered", says Anjali Dewan (Chapter 1). "The goal of environmental sustainability is to minimize environmental degradation and to halt and reverse the processes they lead to." This

Chapter deals with an approach to sustainable development emphasising the need for new conceptions of global development that take cognizance of the fact that social and environmental problems are interconnected. Sustainable Development, Dewan emphasizes, is about a more inclusive society, which provides for better protection of the environment and use of natural resources and shares the benefits of economic growth as widely as possible.

Trauma of the Hot Planet

Global warming is one of the greatest environmental, social and economic threats facing the planet (Chapter 2). Kush Kalra and Amit Kumar are of the opinion that the warming of the climate system is unequivocal, as is now evident from observations of increase in global average air and ocean temperatures, widespread melting of snow and ice, and rising global mean sea level. They quote Stellina Jolly and Varun Bajaj: "While the world's climate has always varied naturally, the vast majority of scientists now believe that rising concentrations of greenhouse gases in the earth's atmosphere, resulting from economic and demographic growth over the last two centuries since the industrial revolution, are overriding this natural variability and leading to irreversible climate change in the global climate system that supports the planet's basic life support functions".

"Foundation" of Global Warming was laid down on the planet with the advent of industrialization in the late nineteenth and early twentieth century. The phenomenon of warming keeps accelerating since then. According to the report of Stern Commission on Climate Change, the planetary temperature has increased one degree Celsius over the last century; a number that might seem small yet has dramatic consequences. The last 24 years have been the hottest in the last century and a half and 2006 was the hottest year since record keeping began in 1856. If no action is taken to reduce carbon emissions, warn Kalra and Kumar, the concentration of atmospheric greenhouse gases could reach double its pre-industrial level within the next 30 years, leading to a temperature rise of at least 2°C.

Environmental Change = Social Change

Environmental change inevitably induces social change. Aparna Rajput and Shiwani Bhatnagar discuss such realities of our times which we are experiencing (Chapter 3). Human society evolves adaptation measures as the environment is flawed by adverse changes. Adaptation to environmental change is not a new concept. Human societies have shown throughout history a strong capacity for adapting to different climates and environmental changes. The Chapter cites many examples; for example, farmers, foresters, civil engineers, and their supporting institutions have been forced to adapt to numerous challenges to overcome adversity or to remove important impediments to sustained productivity.

Human society and the natural environment, nevertheless, are not entirely protected against, nor perfectly adapted to, current climate variability and extreme weather events. Climate instability is resulting in substantial economic losses which are indicative of the fact that the society is vulnerable and that adaptation has not been sufficient to offset damages associated with current variations in climatic conditions. Human-induced climate change, Rajput and Bhatnagar relate, represents a new challenge, and may require adaptation approaches to changes that are potentially larger and faster than past experiences with recorded natural climatic variability.

Climate in Politics and Politics in Climate

Harmfulness of greenhouse gases and their long-term effect on developing countries presses an urgent need to look into the issues emerging out of current scenario of global warming. Shreena Keswani, Deepali Joshi and Anjali Adhikari focus on the politicized issues of climate change (Chapter 4). Biosphere is indivisible. Climate system too cannot be divided. But politics attempts to divide and disintegrate everything. A kind of political squabbling is going on over carbon emission issues the world over. As a result, the emission-relating issues are not getting disentangled. "There is a huge contrast between developed/industrialized nations and poorer developing countries in greenhouse emissions, as well as the reasons for those emissions. Industrialized countries have not provided much help in many areas such as effective emission reductions and stalling on developing country commitments, or opposing the Kyoto Protocol itself", the learned authors claim. "Many scientists say greenhouse emissions need to peak sooner rather than later, so each year seems like wasted time. This big lurking challenge is still on head but we are not convinced to become united and unanimous for the action against global warming."

Keswani, Joshi and Adhikari express their concern that despite a number of meetings, conferences, workshops, seminars, and debates, the challenges before humanity appear to be increasing and hope for a change in the climate change appears to become a distant dream; rather a dream never to come true.

Climate in politics is good and offers some solutions. Politics in climate is bad and it does only make mess out of the climate issues. Climate in politics is acceptable. Politics in climate should be dismissed.

Women at Climate War Front

Women around the world play a key role in the protection of biological diversity through their various responsibilities. Hema Bhatt and Savita Rathour are very right in expressing the notion—"Global warming is not gender-blind". In the adversities generated by climate instability both men and women do not suffer equally (Chapter 5). Facts speak louder than heard: Women are more likely to be hit harder by climate change than men as they have more direct contacts with environmental issues due to their social and family roles and the simple fact that a majority—as much as 70 per cent—of the world's poor are women. Presenting a clue, the authors write that women are especially vulnerable to climate change effects, making up a shocking 80 per cent of climate refugees.

In the war waged by climate, women have to fight at fronts. Women have different set of knowledge and values and their weapons to fight against climate change and vagaries of nature. Bhatt and Rathour call for large-scale women participation and integration into the planning, development, and execution of climate adaptation strategies. Control of population, waste management and alternative energy sources, according to the authors, can contribute to make life of women better in terms of health and work productivity along with a boost in the processes leading to mitigating the climate change.

The effects, or impacts, of climate change may be physical, ecological, social or economic, of which the social ones are the most difficult to observe and have a long-term effect on the general well-being of the society. Shalini and B. Singh highlight another interesting but concealed issue of climate change—the impact of climate change on nutritional status of pregnant women (Chapter 6). This is a case of a drought-hit district of Uttar Pradesh State of India. A majority of pregnant women (90%) has been found suffering from anaemia. Not only this, the children they gave birth to were underweight at the time of birth and prone to diseases due to low level of immunity. Child mortality rate and maternal mortality rate are also likely to be higher amongst the women of weaker sections of society.

Discrimination against women is clearly visible in terms of inadequate and less nutritive diet given to them, resulting in malnutrition and nutrient-deficient diseases. Adverse climate adds fuel to fire. In every adverse situation, including climate change, women have to suffer more than their counterparts. This is the reality of our society. Shalini and Singh underline this fact highlighting results of a research work carried out in a city of Uttar Pradesh.

In the Eye of the Storm: The Psychology of Global Warming

Understanding the fundamental needs of humans is essential if we are to develop strategies for more sustainable development. Areas of contention frequently arise during such discussions at various forums since the goals of each nation, group, or individual which are based upon their needs are often at odds with each other. Ali Seid Mohammed and Uma Melkania discuss the psychology of global warming/climate change (Chapter 7). The issues are deliberated at a number of platforms. UN institutes and committees discuss the pertinent issues but the ball remains in the court of the polluters. Huge amount of money is wasted in the climate summits (also extra emission is added to the environment on account of huge amount of oil to be burnt on extra transport involved), but that goes in vain.

"Interestingly, both the Kyoto Declaration and the WCED view human needs to include the physiological and psychological consequences", the authors inform. They further suggest that by understanding the psychology of human behaviour, it will be possible to develop strategies/programmes that will not only be more effective in communicating the vision of sustainable development to the public, but will also assist society in adopting new behaviours that will help realize global warming threats.

Forests as Human Security Base

Indian culture is believed to be rooted in forests—the *Aranya* culture. But today the state of forests in the country is pitiable. We have only approximately 24 per cent forest cover in the country. Our *Aranya* culture today appears merely a myth. However, tribal populations in many areas of India still live close to the forests and build their livelihoods around the forests. Forests are not only a source of ecological and environmental security, but also a base of food and nutrition security. They serve as a rich repository of a variety of wild foods (fruits, tubers, edible flowers, buds, etc.) which provide security base for tribal populations. Rita Singh Raghuvanshi and Nandana Kumari have investigated a variety of nutritious foods occurring in the forests of Jharkhand (Chapter 8). Diversity of these foods and their nutritive values as well as their utilization by the inhabiting populations has been beautifully described.

Deforestation for agriculture and for other purposes and agriculture itself are one of the prime causes of carbon emissions. Preservation, conservation and augmentation of forests for producing a variety of foods are ecologically sound ways to ensure food and nutrition security.

Community-based Eco-Tourism

Tourism is perhaps the largest industry of the world. It is often regarded as an industry without chimneys. But it has lot of pollution in its fold. It takes heavy toll of natural areas, involves construction activities including roads in the pristine areas of natural beauty, and also cultivates some kind of cultural pollution. However, tourism in today's world cannot be neglected. Bipin Kumar, Dinesh Bhatt, Bitapi C. Sinha and Vir Singh have presented a case of community-based eco-tourism centred around the Jhilmil Jheel in Haridwar district in Uttarakhand (Chapter 9).

"Community-based eco-tourism (CBET) has become a popular tool for biodiversity conservation; based on the principle that biodiversity must pay for itself by generating economic benefits, particularly for local people", argue the authors. "It is a style of tourism that can be used as a tool for conservation of natural areas, and development tool that can be ecologically, economically, and socially sustainable." The study of Kumar and others highlights the potentials of community-based eco-tourism at Jhilmil Jheel conservation reserve. It was declared as conservation reserve in 2006 by the Government of India, to conserve swamp deer and other life forms. It has the excellent habitat of swamp deer and other wild animals.

Joint Forest Management

Deforestation triggered environmental crises and it is a typical example from where the story of environmental degradation begins. Neelam Bhardwaj, Gaurav Papnai and Y.D. Mishra warn the humanity of losing forests at an alarming rate (Chapter 10). We need to protect our remaining forest cover as well as regenerate and sustain this ecological wealth on Earth. "Joint Forest Management (JFM) is a participatory approach to forest management that strengthens communities' capacity to build vibrant local economies while protecting and enhancing their local forest ecosystems by integrating ecological, social and economic components into cohesive approaches to forestry issue", the authors argue. "JFM approaches give local residents both the opportunity and the responsibility to manage their natural resources in an effective way and to enjoy the benefits of that responsibility."

Many Joint Forest Management cases all around the country would reveal that it can help increase income and generate meaningful employment opportunities for local people. It directly protects the environment and maintains the sustainability of the forest reserve. Bhardwaj and her colleagues try to reveal that many factors and problems like non-cooperation of forest officers and villagers, corruption, lack of funding, lack of villagers' participation, etc. are hindering the growth and the objectives of JFM. The development and sustainability of the environment and the forest reserve depends on the proper working of these JFMs.

Sustainable Agriculture : Solution to Many Problems

Food production is a core process the entire humanity revolves around. Agriculture, however, is costing the biosphere and is one of the major reasons of breaking down ecological integrity of the planet. It is not agriculture *per se*, but the kind of unsustainable agriculture we are so much dependent upon for our survival and progress. The conventional, or the Green Revolution type, agriculture demands clearing of forest areas, guzzles lot of water, and leaves a trail of dreaded pollution resulting in poisoning of nature's food chains, ozone depletion, and slowly but steadily transforms the green lands into unproductive deserts.

"Sustainable agricultural approaches help farmers improve local food production with low-cost, readily available technologies and inputs, without causing environmental damage", Kavita Bhatnagar and Shiwani Bhatnagar present their views (Chapter 11). "Sustainable agriculture practices can also mitigate climate change. Organic agriculture, for example, uses less fossil fuel based inputs and has a better carbon footprint than standard agricultural practices. It is thus clear that agriculture needs to undergo a radical overhaul to become more sustainable."

Nanoscience and Nanotechnology : Emerging Opportunities

Nanoscience and nanotechnology are providing new solutions and opportunities to ensure a sustainable energy and environment for the future. Chitra Singh, U.P.S. Bhardwaj and R.P. Singh present many ideas about this emerging area of environmental conservation. "Manufactured Nanomaterials (MN) are attractive to solve numerous energy and environmental issues", assure the authors (Chapter 12). "Nanotechnology offers major benefits to humankind. However, there is growing concern regarding the potentially adverse interactions of engineered nanoparticles and devices on environmental and human health."

Singh, Bhardwaj and Singh stress the need to evaluate the impact of any nanoparticle before it is taken to people who are going to use it or before discharging to environment. This study provides a better understanding of factors governing risks from nanoparticles and current strategies for protecting environmental and public health.

Turning Adversity into Benefits

Liberalisation, Privatization and Globalization (LPG) processes drive the whole world towards a market place. 'Global village' is a metaphor for our contemporary world. But we should better replace this metaphor with 'Global market place'. Yes, today's world of ours is truly a marketplace where everything that comes in your way is price-tagged. Nisha Singh, Deepika Singh, Pratibha Singh, Ashima Srivastava and Roli Verma raise some issues relating to green marketing and present a conceptual approach to solve environmental problems (Chapter 13).

"In today's business world, global and local environmental issues play an important role in marketing scenario", the learned authors go on elaborating. "It is the process of selling products or offering services on the basis of their environmental advantages. The product itself can be environment-friendly or it can be made in an environment-friendly way." The authors are of the opinion that the "green philosophy" would cure many of environmental ills prevailing in our times.

Environmental Psychology to Reveal the Potential of Environmental Goodness

The entire globe has awakened against environmental disasters. However, environmental degradation is going on unabatedly despite this unprecedented awakening of masses. This is out of greed of a few people. "From the capitalist point of view, which calls for maximization of economic profits that can come only through overexploitation of nature, environmental issues are hardly of any value", write Govind Singh Kushwaha and Vir Singh (Chapter 14). "However, principles of sustainable development, that emanate from natural human-nature relationships and that are also embedded deep into human psyche strongly advocate for environment-development reconciliation."

Environmental psychology provides systematic account of human-nature relationship and attempts to promote this relationship and can also promote processes leading to sustainable development. This article reviews the many attributes of environmental psychology.

Community Forests—Cemetery of Carbon

Van Panchayat forests in Uttarakhand have tremendous potential to prove a burial ground for carbon. Vardan Singh Rawat, Ashish Tewari and Y.S. Rawat have conducted a study in which they have estimated carbon stock and carbon sequestration rate in community managed forests (Chapter 15). Their findings indicate that the community managed forests may act as role model in mitigating climate changes and play a major role as carbon sink. "We must encourage the community forests and the people living in and around these forests for saving their forests", they suggest adding that determining the level of carbon stock pools in different components has become a concern of governments, businesses, and many organizations. It is a must to incorporate climate change consideration in forest sector especially Van Panchayat forests for long-term planning process.

A large chunk of forests in the Himalayan mountains are in the common property regime. These forests especially need attention of management. Village-based organizations, including Panchayat bodies, should be encouraged to protect and regenerate forest lands and contribute to enhance the processes of carbon sequestration. Community forests can really make a difference in our national planning having bearing on climate change mitigation.

Educating for Pollution-free Household Atmosphere

In the mountains of Uttarakhand, as also elsewhere in the Himalayan region, women play a distinctive role. All economic activities of a household revolve round women farmers. But the indoor pollution owing to the use of conventional energy source for cooking food they continuously suffer from numerous health problems. Getting rid of the indoor pollution is critical for a healthy environment and happy and healthy life of woman farmers. Women are not adequately educated and trained so that they could be able to respond to indoor pollution becoming havoc for their lives. Kiran Yadav, Neelam Bhardwaj and Uma Melkania provide interesting suggestions and measures to get rid of the indoor pollution which continues to be part and parcel of hill women's lives (Chapter 16).

The learned authors present a case of educating women in a unique way, which not only enables women to make their kitchens pollution free but is also full of recreation that would be of great help in emotional and psychological development of all family members.

Food for Thought: Food Habits affecting Environmental Health

Food habits affect environmental health. Should we improve the health of our ecosystems, we should look into our food habits. Junk foods or fast foods are now part and parcel of the lifestyles of our young generation. But these foods are no good either to human health or environment. These foods rather invite a host of problems to both environment and society. Rita Patnaik and Gita Singh bring to the fore numerous fast food-related risks to the health of people (Chapter 17). "Promotion of such foods all over the world is opening Pandora's box of problems which can cost humanity", warn the authors. "Slow food—an antithesis of fast food—which are nutritious and health-friendly and part of an ecologically sound agricultural system and of our rich traditions, are good for the society as well as for the ecosystem health. This article provides several attributes of the slow food. "Slow food unites the pleasure of food with responsibility, sustainability and harmony with nature", says Carlo Petrini.

Organic foods being produced and promoted by Navdanya of Dr. Vandana Shiva, Slow Food concept steadily capturing attention of people worldwide, traditional vegetarian foods, *baranaaja* culture of mountains, etc., present a variety of foods for thought. A happy and nutritious India must be the goal of our march with right kind of food habits.

Environmental Resistant Movement

Environment resistant movements in a democracy are very auspicious. It has been an experience in India that it is people, not the government, who generated lot of environmental awareness and also led crucial environmental conservation programmes. Chipko Movement, Anti-Tehri Dam Movement, Ganga Bachao Andolan, Narmada Bachao Andolan, Appiko Chaluvali, etc., are the names of a few environmental movements Indian masses observed in recent years and which have become exemplars for the entire humanity fast awakening for environmental issues. D. Jayadeve Gowda has covered one such environmental resistant movement of agrarian community going on against Bagur-Navile tunnel project in Hassan district of Indian State of Karnataka for last several decades (Chapter 18). Farmers have protested against the project for a long time without getting any substantial support from political leaders and administrative machineries. The planning officers have designed and constructed about 200 ft depth tunnel from the soil surface without forecasting knowledge.

The Government has implemented the project against the interest of the agriculturists and the farmers of Bagur Navile area. D. Jayadeve Gowda discusses the role of agrarian community in resisting the project and major points of objection raised by them. He also elaborates the ideas of agitation adopted by the agrarian community and local group against the controversial project. He concludes how the resistant movements were suppressed and misguided by the political administrators of Karnataka State and Hassan district. He also suggests that acceptable relief should be given to the affected groups of people by taking into consideration the confidence of civil society organization.

Climate Change, Community and Communication

Climate change cannot be dealt with a single formula, but by a complex set of formulae. The climate phenomenon is neither simple, nor linear type. It is a complex and not to be disentangled so easily. "Climate change and agriculture are interrelated processes, both of which take place on a global scale", writes Arpita Sharma. "Worldwide, farming is a significant source of greenhouse gas emissions, accounting for a sizable share of the estimated 20 per cent of total emissions that result from land use." Agriculture's vulnerability to climate change and the grave consequences—this implies for the world's poor and hungry (Chapter 19).

In the rural area the farmers have less knowledge about the negative effects of climate change on agriculture. Climate change is posing real threats to food, water and human security as well as to the gains made in development and poverty reduction. Arpita Sharma suggests that there is an urgent need to develop improved strategies to adapt to impacts and to raise awareness about how climate change is affecting peoples' lives and livelihoods.

As the learned author suggests based on her research experiences, an effective extension, education and communication services are probably some of the key strategies for sustaining agricultural growth, strengthening food security and combating hunger and malnutrition. Community radio can play an important role in facilitating the sharing of experience, information and knowledge with and between those who are coping with these impacts most directly. It can also serve as a forum to call for national and international action to support communities dealing with current and future impacts.

Part I

LIVING IN A CLIMATE OF UNCERTAINTY
Malady of the Development and New Challenges facing Humanity

1

Environment and Sustainable Development : Paradigm Shift with a Vision for Future

Anjali Dewan

Introduction

Sustainable development is a socio-ecological process characterized by the fulfilment of human needs while maintaining the quality of the natural environment indefinitely. The linkage between environment and development was globally recognized in 1980, when the International Union for the Conservation of Nature published the 'World Conservation Strategy' and used the term 'sustainable development'. The concept came into general usage following publication of the 1987 report of the Brundtland Commission—formally, the World Commission on Environment and Development set up by the United Nations General Assembly. This became the most often quoted definition of sustainable development as development that 'meets the needs of the present generation without compromising the ability of future generations to meet their own needs'. It refers to the wise use of resources within a framework in which environmental, economic and social factors are integrated. It is about maintaining and improving the quality of life while safeguarding the quality of life of generations to come. The components of biological diversity have to be utilized in such a manner that does not interfere with the natural functioning of ecological process and life-support systems. It is a collection of methods to create and sustain development which seeks to relieve poverty, create equitable standards of living, satisfy the basic needs of people.

For India, a large country both populated and poor, to develop in an environmentally sustainable development is not an option but a requirement. On the one hand, India is faced with environmental degradation from poverty and population pressures, and on the other, from pollution from increased activities due to economic growth and the consequent changing consumption patterns. While the poor depend on the environment for their livelihood, the process of economic development relies on using natural resources to produce goods and services. The waste generated from consuming and producing these goods and services are in turn released back into the environment impacting it. The environment provides security for present and future generations, the health of the environment is closely connected with the health of humans, and it is economically beneficial for countries to prevent environmental degradation. The challenge, therefore, in making development compatible with the environment is to restructure the economic system in a way that it will not destroy the environment as economic progress continues.

Sustainable development does not focus solely on environmental issues. The 2005 World Summit Outcome Document, refer to the interdependent and mutually reinforcing pillars of sustainable development as economic development, social development, and environmental protection. Green development is generally differentiated from sustainable development in that green development prioritizes what its proponents consider to be environmental sustainability over economic and cultural considerations. Proponents of Sustainable Development argue that it provides a context in which to improve overall sustainability where cutting edge Green development is unattainable. For example, a cutting edge treatment plant with extremely high maintenance costs may not be sustainable in regions of the world with less financial resources. An environmentally ideal plant that is shut down due to bankruptcy is obviously less sustainable than one that is maintainable by the indigenous community, even if it is somewhat less effective from an environmental standpoint. Some research activities start from this definition to argue that the environment is a combination of nature and culture.

Environmental sustainability is defined as the ability of the environment to continue to function properly indefinitely. This

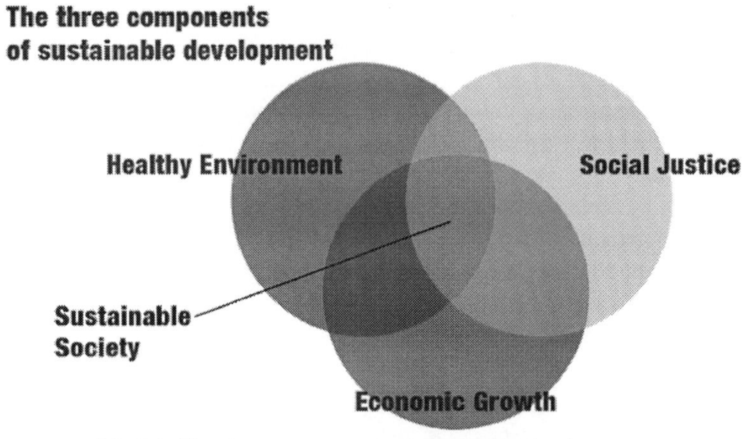

Fig 1.1: The three components of sustainable development.

involves meeting the present needs of humans without endangering the welfare of future generations. The goal of environmental sustainability is to minimize environmental degradation, and to halt and reverse the processes they lead to. An unsustainable situation occurs when the sum total of nature's resources is used up faster than it can be replaced. Sustainability requires that human activity only uses nature's resources at a rate at which they can be replenished naturally. The long-term result of environmental degradation would be local environments that are no longer able to sustain human populations to any degree. Such degradation on a global scale could imply extinction for humanity.

The gross mismanagement of our planet has much to do with an inequitable distribution of the benefits of development. Perpetuating this inequity can only mean a continuing drawdown on the world's natural resources and the environment. After a century of unprecedented growth, marked by scientific and technological triumphs that would have been unthinkable a century ago, there have never been so many poors, illiterate and unemployed people in the world, and their number is growing. Close to a billion people live in poverty and squalor, a situation that leaves them little choice but to go on undermining the conditions of life itself, the environment and the natural resource base.

The key socio-economic pressures on the environment such as population, poverty, and illiteracy on the one hand, and economic growth, changing consumption patterns, structure of GDP and trends of urbanization on the other. The state of India's environment with respect to air, water and land as is well-known, India is set to replace China as the most populous country in the world with 1.5 billion people by 2040. Since 1951, the population has grown from 361 million to 1 billion making every sixth person in the world Indian. As India has only 2.4 per cent of the world's total landmass, the population density is at 324 persons per square kilometre (Census 2001) and pressures on the environment are correspondingly high.

Issues and Challenges

It is widely accepted that meeting the needs of the future depends on how well India and the rest of the world balances their social, economic and environmental goals when making decisions today. Research carried out under consecutive research Framework Programmes has addressed sustainable development, in particular, and environmental issues, in general, in a number of different contexts. The challenge, therefore, in making development compatible with the environment is to restructure the economic system in a way that it will not destroy the environment as economic progress continues. Broadly speaking, the challenges facing us today are as follows :

- The ways in which greenhouse gas emissions and atmospheric pollutants affect climate, ozone depletion, soil and forests.
- The water cycle and soil-related aspects, especially looking at the impact of climate change on the hydro-cycle, interactions between land, ocean and atmosphere, distribution of groundwater and surface water, fresh water and wetland ecosystems, management strategies for water (availability, demand, vulnerability, quality, etc.)
- Understanding marine and terrestrial biodiversity to minimize the negative impact of human activities, relationships between society, the economy and biodiversity habitats, forecasting potential changes, studying the structures and mechanisms while assessing and managing the risks to biodiversity and ecosystems, as well as looking into

conservation, mitigation and rehabilitation options for marine and terrestrial biodiversity.

- Mechanisms of Natural Disasters and Desertification such as seismic and volcanic activity and how these tie in with climate change, land and soil degradation and studying what science can do to help long-term forecasting of hydrogeological hazards, natural hazard monitoring, mapping and management, and to improve disaster preparedness and mitigation strategies.
- Strategies for sustainable land management, including integrated coastal zone management and assessment tools for measuring the impacts of agriculture and forestry on the environment, while looking at new technologies and integrated approaches to make these sectors more environmentally friendly.
- Systematic operational forecasting and modelling of climate change (atmospheric, terrestrial, marine-based) to improve understanding of extreme events, and to use data coming from the Global Monitoring for the Environment and Security (GMES) programme and other Global Observation Systems for the climate, oceans and terrestrial more effectively.
- Cross-cutting issues, such as putting together sustainable development concepts and tools for measuring the sustainability dimension on relevant policies towards sustainable development in a very thorough, practical and meaningful manner.

Certain important factors which have an impact on the environment and sustainable development are as follows:

Socio-economic Pressures on the Environment

There is pressure on the environment such as population, poverty, illiteracy on the one hand and economic growth, consumption patterns and urbanization on the other.

Population

As is well known, India is set to replace China as the most populous country in the world with 1.5 billion people by 2040. Since 1951, the population has grown from 361 million to 1 billion making every sixth person in the world Indian. As India has only 2.4 per cent of the world's total landmass, the population density is at 324 persons per square kilometre (Census 2001). While the growth rate of the population has declined over the years, the population in India continues to be high on account of the large base of the population in the reproductive age group. The higher contribution to fertility due to the unmet need for contraception is estimated to be 20 per cent. High fertility rates are seen also due to high infant mortality rates, which are exacerbated by the lack of sanitation and health care facilities. In addition, 50 per cent of girls marry below the age of 18, resulting in a typical reproductive pattern 'too early, too frequent, too many' (GoI, 2001). The significance of population pressure is great as it impedes both development and environmental health.

Literacy

It is well known that literacy is important for just about every aspect of development and the environment is no exception. For instance, women's literacy reduces fertility rates and thus reduces population pressures. Literacy is essential to reducing poverty as it provides people with more opportunities to earn as well as increases their productivity and the ability to earn. It also allows people to become more receptive towards alternative technologies and environmental management. People's ability to manage environmental resources like water, soil, and forests improves with literacy. Thus, literacy helps reduce pollution as the added awareness allows people to link the health of the environment with their own health and well-being. In India, literacy rates are growing steadily but they are still far from adequate.

Poverty

Clearly, one of India's biggest problems is poverty. Fortunately, despite the growth of the population and the low level of economic development at the time of Independence in 1947, India has made significant progress in poverty reduction. The percentage of people below poverty line has come down significantly. Yet, a large number of persons remain below the poverty line. The high incidence of poverty that still prevails in India underlines the need for rapid economic development to create more remunerative employment opportunities and to invest in social infrastructure of health and education. The pollution that the poor generate is of a different kind than generated by the rich. The poor from lack of alternatives rely intensively on the environment for their needs; its degradation affects their livelihood immensely. They thus become both victims and agents of environmental degradation as they both need and often exploit the environment to survive but suffer the most when it degrades. On the other hand, the poor are also victims of pollution. When water sources are polluted, the poor women have to walk further and spend considerable time fetching it. Lack of resources such as clean fuels and safe drinking water results in back-breaking work to fetch them along with health consequences due to water borne diseases and respiratory, eye diseases and reproductive health. The rural population spends 15 billion workdays gathering water and fuels every year suffering health consequences. This should be seen as a tremendous national loss. How can India achieve 8 per cent economic growth on a sustainable basis if nearly 35 per cent of the population undergoes hardship and drudgery doing daily chores.

Urbanization and Transportation

As countries industrialize one sees cities grow and traffic increase. This may reflect rapid economic development and industrialization but also results in high levels of energy consumption and emissions. The progress of urbanization is relatively slow in India compared to other developing countries and the bulk of the population continues to live in rural areas. The urban population increased from 19 per cent of the total population in 1965 to 28 per cent in 2000. It is expected that more than 50 per cent of population in India will reside in urban areas by 2025. The number of urban agglomerations/cities with populations of over a million has increased from 5 in 1951 to 9 in 1971 and 23 in 1991 to 35 in 2001. The increase in the requirements of urban transportation in India over past few decades has outstripped the development of urban infrastructure. This has led to congestion of cities, proliferation of private vehicles, increased energy usage and increased pollution. The situation got worse due to shift of modal split away from railways towards road transport and from public to private transport due to inadequate capacity expansion of public transport leading to rising emission levels of pollutants and increased energy usage due to proliferation of private vehicles. Transport is a critical infrastructure for both intercity and urban development. When low-income groups do not have access to affordable transportation system, this imposes hardships and impacts their quality of life. Their time and energy is wasted in commuting, making them inefficient and thus sending them in a vicious circle of poverty and inefficiency. Efficient transport system is a critical infrastructure for the cities for economic productivity and quality of life. Transportation sector is responsible for a large share of air pollutants ranging from respirable particulate matter, nitrogen oxides, carbon monoxide, ozone and greenhouse gases etc. Therefore, in order to develop sustainably greater emphasis needs to be placed on developing good public transport systems.

Water

Water pollution is a major cause of concern in India because in addition to causing ecosystem damage it adversely affects health and thereby impairs economic productivity of people. The current condition of water resources requires that urgent action be taken. For instance, 90 per cent of India's surface water resources are polluted to the extent that they are not fit for bathing. Also, about 200 million people do not have access to safe drinking water and utilizable water per capita is decreasing. The socio-economic costs of water pollution are extremely high : 1.5 million children under 5 years age die each year due to water related diseases, 200 million man days of work are lost each year and the country loses about Rs. 36,000 crore each year due to water related diseases.

Agriculture Industry Domestic Pollution

The three major contributors towards water pollution are the domestic sector, the industrial sector and the agricultural sector. 75 per cent of the effluents by volume are from the domestic sector. This is because only 20 per cent and 2 per cent of wastewater in Class I and Class II cities respectively is treated. Meanwhile, only 3.15 per cent of rural population has access to sanitation services and 115 million homes have no access to toilets of any type. In the industrial sector only 59 per cent of large and medium industries had adequate effluent treatment in 1995. In the agricultural sector, fertilizer use increased from 7.7 MT in 1984 to 13.4 MT in 1996 and pesticide use increased from 24 *TT* in

1971 to 85 TT in 1995. One of the reasons why environmental standards are ignored is because they are seen to be expensive. However, when one calculates the cost of lowering pollution one must compare it to the health and economic benefit from abating pollution. For example, Rs. 460 billion is needed to construct toilets in 115 million homes, wastewater treatment in 3,696 cities/towns would cost Rs. 180-600 billion depending on technology, pollution abatement in industries would cost Rs. 140 billion (about 1.2% of total annual turnover). However, the loss from human health damages due to sanitation and water pollution is rupees 360 billion per year.

Air

Contrary to popular belief, air pollution is not an urban problem alone. While in cities suspended particulate matter, sulphur dioxide and nitrogen oxide levels are much higher than permissible limits, in rural areas indoor pollution kills half a million prematurely every year. In urban areas the transport sector is the highest contributors to air pollution, although refuse burning and Industrial pollutants are also big contributors. Suspended particulate matter and respirable suspended particulate matter are at very high levels in both residential and industrial areas. In rural areas, air pollution is primarily produced indoor from the use of firewood and other unclean sources of cooking fuel. Since 72 per cent of India lives in rural areas, this form of air pollution has significant impact on the population as a whole. Indoor pollution is the cause of 0.41 to 0.57 million premature deaths per year.

The Need for a New Development Paradigm

The need for a new development paradigm was widely recognized by the mid-1980s. By then, the United Nations had already declared the second and third of its Development Decades a failure, especially in their inability to halt the cycle of poverty that existed in the world's poorest and slowest developing countries. Further, the conditioned loan policies of the World Bank and the International Monetary Fund in combination with patterns of development assistance provided through various bilateral and other arrangements succeeded in plunging already impoverished countries into even deeper debt. Thus, the combination of high debt, slow economic growth, and rapidly increasing poverty enhanced the profound levels of human suffering that already existed in many of these countries. Groundbreaking events, such as the United Nations-backed Rio Earth Summit on Environment and Development in 1992, and the Johannesburg Summit on Sustainable Development a decade later, show that policy-makers recognized the grave challenges that lie ahead in preserving the planet's fragile environment. One of today's environmental challenges is to find ways of strengthening the scientific and socio-economic perspectives to help authorities make decisions and produce sustainable development strategies.

Sustainable Development Strategies

The Government has emphasized the need for a coordinated approach to sustainable development. Its efforts to advance sustainable development in India and around the world are also evidenced by its action on climate change, human health and the environment; its support for the building and maintenance of green infrastructure in and its recognition of the unique relationship tribal people have with the natural environment. In addition, the Government has also recently made a commitment to regularly update strategies, which will ensure progress. Some of the common themes that have emerged in this regard include :

- Enhancing the capacity of Indians to act and make decisions in support of sustainable development; strengthening and improving access to knowledge in support of integrated decision-making.
- Raising the level of understanding/awareness of sustainable development issues; providing the necessary tools to policy-makers to a better advanced sustainable development.
- Building on existing partnerships and creating new ones; providing leadership in areas such as greening of operations; fostering innovation and technology and promoting the adoption of sustainable development policies and practices in India and abroad.
- A coordinated approach, which outlines best practices for departments to adopt in procurement, waste management, water conservation, energy efficiency, vehicle fleet

management, land use management and human resource management needs to be prepared.
- Eco-efficient business practices that conserve energy, eliminate use of toxic materials, and reduce or reuse resources, innovative eco-efficient practices, tools, technologies and products can increase productivity while improving environmental performance.

The quality of life of Indians is intrinsically linked to the health and well-being of their communities. A sustainable community is resilient to change and has a vision for the future. However, to become sustainable, communities face complex environmental, economic and social challenges and require a coordinated approach to information, support and expertise.

Sustainable Development and Healthy Indians

Human health is dependent on a wide range of factors including the natural and built environments, and social, cultural and economic conditions. Certain populations like children, pregnant women and lactating mothers, the elderly and tribal people may be particularly vulnerable. A meeting with representatives from a range of sectors needs to be initiated to discuss the implications of human health when making decisions for sustainable development. The intention is to continue this dialogue whereby departments will explore information needs, strategies for action and approaches to engage the public in order to protect the health and overall quality of life of our people.

Social and Cultural Aspects of Sustainable Development

There is growing recognition that the concept of sustainable development is multifaceted with links to social and cultural aspects, such as poverty, distribution of wealth, diversity, equity, gender and health. These not only may affect the environment, but also influence and shape economic behaviour, consumption patterns, and the longer term viability and sustainability of our society. We are at an early stage of understanding the nature and implications of the social and cultural dimensions of sustainable development. Accordingly, various governmental and private agencies will be working together to better understand this area and the relationships involved by undertaking exploratory research and providing social policy perspectives.

Law Enforcement and Other Controls

- *Consolidate Pollution Control Laws*: The environment has to be addressed holistically as different media like air, water and soil are interconnected and pollution once generated, can shift in space and time. That is, pollution can be diluted in more water or air or converted into different forms of pollutants like burning solid waste and causing air pollution or pollution laws must be consolidated to address this issue.
- *Introduce Full Liability Laws*: Industries and other polluters should be made fully liable for their pollution. After the Bhopal tragedy this should have been obvious, yet the Bhopal victims have not got adequate compensation.
- *Make Clean Technologies Mandatory in New Industries*: The current industry growth rates are around 7-8 per cent. In this light, industries being established now will comprise 50 per cent of the market in 9 years (Jayaraman, 2001). Cleaner technologies should be made mandatory in such industries so as to reduce pollution in the coming years. Industries are less willing to change their techniques once they are established. Hence, they should be made to take on cleaner technologies right from the start. The technology for small-scale industries can be improved by requiring large-scale capital goods manufacturers to produce less polluting equipment.
- *Make Functioning Treatment Facilities Mandatory*: Often, due to mandatory requirements, effluent treatment plants are set up but not operated to save operating costs. Therefore, laws should provide for monitoring operation of these facilities.
- *Require Environmental Audits for Industry*: Environmental impact assessments are often made at the start of the project but prescribed environmental management practices are often forgotten once the project is underway. Regular audits would act as self-monitoring and enhance compliance to standards.
- *Provide Effective Right to Information*: If people have the information about what their neighbouring industry pollutes,

they would generate pressures for abatement and treatment. Full liability laws need to be complemented with right to information.

Role of Pollution Boards

Central and State pollution control boards need to be strengthened technically by installing modern equipment and training, financially with larger budgets to hire better staff and facilities to monitor the pollution, and politically by keeping them independent and free from political influence. This has to be done at all levels, e.g., at municipalities and Gram Panchayats levels to control pollution. Suggestions have also been made that industries should be roped into monitor pollution and assist new industries in choosing environmentally friendly technologies. Involving private sector in water supply and sewage treatment in urban areas may prove to be beneficial.

Role of Centre for Environmental Education (CEE)

The primary objective of CEE is to improve public awareness and understanding of the environment with a view to promoting the conservation and sustainable use of nature and natural resources, leading to a better environment and a better quality of life. It develops innovative programmes and educational material, builds capacity in the field of education for sustainable development. It undertakes demonstration projects in education, communication and development that endorses attitudes, strategies and is committed to ensuring that due recognition is given to the role of education in the promotion of sustainable development. The activities of CEE and programmes have been guided by certain strategies for maximisation of quality, effectiveness and impact. These are as follows :

- *Adaptability* to different geographic, cultural, social and economic contexts.
- *Partnerships* utilizing complementary strengths of other organizations to avoid duplication of effort and the network effectively for synergistic convergence of ideas and goals.
- *Building synergies* among Government, NGOs and CEE for comprehensive impact.

- *Identifying key entry points* for different thrust areas and key targets for initiating and consolidating gains to achieve a multiplier effect.
- *Facilitating networks* to local, national and regional levels through a number of tools such as dialogues, directories, newsletters, etc.
- Working to *develop a cadre of professionals* in order to improve and strengthen professionalism in the field of Education for Sustainable Development (ESD) by capacity building individuals who in turn would infuse this professionalism into the organizations where they work to the mutual advantage of both those organizations and CEE.
- Ensuring *quality control and excellence* in the production of all material through in-house infrastructural support like the studios, workshops, printing facilities.

The Path Forward

As the sustainable development journey continues, the various departments will continue to learn and improve. There will be further efforts to coordinate between departments as we move to the implementation of the updated strategies. This will include efforts related to communication and awareness building amongst our countrymen, science and technology development and staff training. The Government will also move forward on the current priority areas and identify opportunities for additional collaboration. The transition to a more sustainable society in the twenty first century will not be achieved by one department or one level of government alone. Moving towards more sustainable development will require the commitment and attention of decision-makers at all levels of society. The Sustainable Development Strategies are an important element of the Government of India's response to this significant challenge. The strategies will identify common goals and actions towards ensuring healthy human and natural environments, supporting emerging governance systems, advancing appropriate use of natural resources, building vibrant communities, and influencing international activities. Sustainable Development is about a more inclusive society, which provides for better protection of the environment and use of natural resources and shares the benefits of economic growth as widely as possible. It is about achieving a better quality of life.

It is not possible for the Government to monitor pollution and the corresponding acts of all industries and individuals. People must be made stakeholders in the environment through awareness campaigns. Industries are sensitive to public pressure. Experience in the West suggests that firms wish to maintain a green profile when citizens are aware of environmental issues. Through generating awareness, the public could directly affect the environmental practices of industries. India has to move forward technologically in an overall context to avoid pollution at source. Modern technologies already developed in the developed countries such as better power plants and cleaner vehicles should be considered through technology transfer for pollution measurements. Technological innovations allow India to develop rapidly in an environmentally sustainable manner. India cannot afford to neglect the environment, if sustainable development is desired. The pressures on environment have to be curtailed by reducing population pressure, increasing literacy, environmental awareness drives and poverty alleviation programmes. The poor have to be turned into agents for environmental restoration by involving them in say forest management, waste management, recycling and so on in a manner that creates incentives for them to use natural resources in sustainable manner. The economic activities must be conducted using environment conserving and resource saving technologies. Managing environment through better urban designs, improvement in transportation infrastructures and creative use of information technologies needs to be considered seriously. The strategy for environmental governance should consist of law enforcement, providing economic incentives, people's participation, institutional reforms and support and technological improvements. With the determination of the government, private sector, NGOs, and people, India can perhaps achieve sustainable development.

REFERENCES

Alger, C.F. 1990. Grass-roots perspectives on global policies for development. *Journal of Peace Research*, Vol. 27 (2), 155-168.

Brandt Commission. 1980. *North-South: A Programme for Survival.* Pan Books, London.

Burek, D. (ed.). 1991. *Encyclopedia of Associations: International Organizations*. New York: Gale Research Inc.

Campfens, H. 1990. Issues in organizing impoverished women in Latin America. *Social Development Issues*, 13(1), pp. 20-43.

Census of India. 1993. *Emerging Trends of Urbanization in India: An Analysis of 1991*. Census of India, Government of India, New Delhi.

Census of India. 2001, *Series* 1, *India*, Paper 1 of 2001.

Central Water Commission (CWC) 1998. *Water Statistics of India*, Delhi.

Economic Survey of Delhi—1999-2000.

Ester, R.J. 1993. Towards Sustainable Development: From Theory to Praxis. *Social Development Issues*, Vol. 15 (3), pp. 1-29.

Jayaraman, K.S. 2001. Indian regulatory system stifles industry growth, *Business and Regulatory News*. Vol. 19, Number 2, pp. 105-106.

Mexico City's Water Supply: Improving the Outlook for Sustainability. 1995.

Midgley, J. 1990. International Social Work: Learning from the Third World. *Social Work*, Vol. 35 (4), pp. 295-301.

Nayak, R.K. and Siddiqui, H.Y. (eds.) 1989. *Social Work and Social Development*, Geetanjali Publishing House, New Delhi.

World Bank. 1993. *World Development Report 1993: Investing in Health*. Oxford University Press, New York.

World Commission on Environment and Development. 1987. *Our Common Future*. Oxford University Press. New York.

World Development Report. 1990. *Global Poverty*. Oxford University Press, New York.

2

Threat of Global Warming : Concern and Challenges

Kush Kalra and *Amit Kumar*

> *The two issues of this Country are the issues of Human Rights and Environment. Healthy environment is also a Human Right.*
> —Justice A.M. Ahmad

Introduction

We are living in an era of momentous global ecological changes and challenges as in the name of the development mankind is disturbing the delicate poise, which the Nature has designed for it. Historically, the humans have revealed frequently a remarkable capability to survive within the restricted wealth of the Earth. As Gandhi's famous pronouncement emphasied, "Earth provides enough to satisfy everyone's need" has become the prime predicament of humanity. Though the environment pollution is not new and it was recognized as a problem in the Ancient Hindu Scriptures as well as at the time of Plato, some 2500 years ago (Hambro, 1994).

The principles and rules, which have evolved in the process of preventing the alarming prospects of ecocide, are called international environmental law. The San Francisco and Bretton Woods Conferences launched history's most prolific period of international institution and law building. On 25th June, 1946, UN Conference on International Organization unanimously adopted the Charter of UN. In July 1944, the UN Monetary and Financial Corporation held at

Bretton Woods, New Hemisphere adopted the IMF and World Bank agreements (Werksman, 1996). The establishment of both UNEP and CSD followed two landmarks global conferences convened by the General Assembly. The international efforts to protect the environment date back to at least 1870 when Switzerland tried to establish a regional agreement to protect the nesting sites of the migratory birds. But the move to institutionalize the law did not get a serious momentum until 1970s. Since then over 200 standards have been adopted covering environmental subjects of shared concern such as endangered species, ozone depletion, climate change, etc. More than two-thirds of these standards, including both the soft and the hard law, have been adopted since the UN Conference on the Human Environment held at Stockholm in 1972. The Rio de Janeiro Conference in 1992, i.e., The Earth Summit I, along with Earth Summit Plus Five held at New York in 1997, and Earth Summit Plus Ten on Sustainable Development held at Johannesburg in 2002, has added new dimensions and pace to international law making. For convenience sake these standards could be divided into two categories, viz., General and Special standards.

General International Environmental Law

The General International Law has taken up the challenges related to threats to global commons, trans-boundary spillovers and disparity in financial and technical capacity, etc. The global commons include biological diversity, ozone layer and the climate change, etc. As the activities of all states could be affected by threats to these resources, therefore, the international law has tried to establish the strategic responses based on exchange of information and common goals. Similarly, the pollutants can spill over boundaries and have impact on the environmental and economic interests of other countries. As all the states seek to move towards a higher level of protection of environment, international law through institutions like World Bank, IMF, UNDP and UNIDO, etc., can provide financial, intellectual and technological resources. Following are some of the important documents of general international environmental law:

Declaration of the Permanent Sovereignty over Natural Resources 1962

This declaration recognizes the permanent sovereignty of the states over their natural resources. It also identifies the need for international cooperation in the field of economic development. However, it reminds the states that such sovereignty should be exercised only for national development and well-being of people.

Stockholm Declaration 1972

This declaration resulted from Stockholm Conference on Human Environment in 1972. This Conference marked a watershed in international relations as it placed the issue of the protection of biosphere on the official agenda. The major contributions of this conference were:

Concept of Sustainable Development

This term was used at the first time in Cocoyoc Declaration on Environment and Development in the early 1970s. But it gained the real inputs during this Conference. The main thrust of this concept is integration of development and environmental imperatives.

Common Agenda

The Conference called upon the Governments and the people to exert common efforts for the preservation and improvement of human environment for the benefit of all the people and for their posterity. For this purpose, it adopted 26 principles as the common agenda of the mankind for the preservation of the environment.

United Nations Environment Programme

Before Stockholm Conference, several specialized and related agencies of UN system were dealing with environment. However, this Conference created a specialized environmental programme as well as UNEP as a focal environmental agency in the UN system. Since then UNEP has been acting as UN's conscience keeper in the field of the greening process.

The World Charter for Nature 1982

General Assembly adopted this Charter with a view to raise awareness that mankind is a part of nature and life depends on uninterrupted functioning of natural system. The state parties were called to protect and preserve nature assimilating the principles of this Charter in their laws and practices.

Brundtland Commission 1983

On the recommendations of the UNEP, the World Commission on Environment and Development was set up by the General Assembly in the year 1983. The influential report of this Commission entitled 'Our Common Future' brought the concept of sustainable development into widespread use. This gave a comprehensive definition to the term sustainable development in the following words :

> "Sustainable development is the development that meets the need of the present without compromising the ability of the future generations to meet their own needs."

The Montreal Protocol on Substances that Deplete the Ozone Layer (1987)

The main objective of this protocol is the purging of ozone-depletion substances at an identical pace regardless of the development status of the country.

Hague Declaration on Environment 1989

This Declaration acknowledged that haphazard human activities on the earth are posing a threat to earth's atmosphere as well as the basic human right to life. This situation calls for crucial, pressing and international solution to this problem. Likewise effective decision-making and enforcement machinery is must.

Earth Summit 1992

This was the largest UN Convention ever held, as 150 Governments participated in it. This Summit inspired and guided by Brundtland Commission report placed the humanity on the path of sustainable development. Some of the important accomplishments of this Summit lie in the following documents :

- *UN Declaration on Environment and Development*: It contains 27 principles defining the rights and responsibilities of the states in this area.
- Agenda 21: It is a comprehensive blue print for worldwide action in the twenty first century to effect shift to sustainable development.

Sustainable Development

Thus, it appears that since 1972, the axiom of the international law has been sustainable development, which has achieved diverse magnitudes in terms of economic expansion, development and environmental safety. Some of the features of sustainable development are :

Inter-generation Equity

The innermost premise of this concept is that each generation has a right to get legacy of biodiversity from the past generations. Moreover, the present generation is under an obligation to preserve such inheritance as well as the renewable resources, such as forests, water and soil, for future generations.

Preservation and Upgrading of Natural Resources

The resources must be conserved and enhanced to meet the necessities of sprouting people. Thus, present cohort should be reticent in their exploitation of natural resources. This idea has achieved widespread international endorsement since the Maltese Proposal at General Assembly of 1967. This proposal stressed that the natural resources, such as sea bed, are not the outgrowth of the labour of present generations only. Hence, these should be exploited with adequate consideration of the rights of the future generations.

Protection of Environment

Without adequate environmental protection, the development is weekend and without development resources will be scarce for needed investments and environmental protection will collapse.

Precautionary Principle

It is based on the theory that it is better to err on the side of prudence and prevent environmental harm, which may indeed become irretrievable. The environmental protection should not only aim at protecting health, property and economic interests but also protect the environment for its own sake. The precautionary duties must not only be activated by the suspicion of tangible danger but also by justified concern or risk potential (*A.P. Pollution Board* v. *Prof. M.V. Nayudu*, 1999). According to Supreme Court of India, this principle means that:

- The government must anticipate, prevent and attack the causes of environment degradation.
- The measures to deal with the threats of serious damage should not be deferred in the name of lack of scientific proof of linking that particular substance or activity, introduced as a result of human intervention, to environmental degradation.
- The burden of proof is on the developer to show that his action is environmentally benign (*Vellore Citizen's Welfare Forum* v. *Union of India, 1995*).

Assimilative Capacity Principle

However, in the Stockholm Conference the spotlight was on Assimilative Capacity Principle. This principle presumed that relevant technical expertise would be available at the time of environmental harm prediction and sufficient time would be available to avoid harm. Nevertheless, by the time of World Charter for Nature 1982, the stress revolved to the precautionary principle. This principle was reaffirmed in Rio Declaration 1992.

U-turn in Burden of Proof

The adoption of the precautionary principle led to reversal of burden of proof in environmental cases. Instead of compelling the persons opposing a particular project, to shoulder the burden of proving its detrimental impacts on the environment, it has shifted the burden of proof, as to the absence of injurious effects of the action proposed, to those intending to disturb the *status quo* (Olso, 1990).

International Negotiations

Climate change emerged on the political agenda in the mid-1980s with the increasing scientific evidence of human interference in the global climate system and with growing public concern about the environment. Because climate change is such a complex and challenging issue, policy-makers need an objective source of information about the causes of climate change, its potential environmental and socio-economic impacts, and possible response options.

Recognizing this, the World Meteorological Organization (WMO) and the United Nations Environment Programme (UNEP) established the Inter-governmental Panel on Climate Change (IPCC) in 1988. The Panel's role is to assess on a comprehensive, objective, open and transparent basis the best available scientific, technical and socio-economic information on climate change from around the world. In the first report in 1990, IPCC concluded that the growing accumulation of human made greenhouse gases in the atmosphere would "enhance the greenhouse effect, resulting in an additional warming of the Earth's surface" by the next century, unless measures were adopted to limit emissions. The UN General Assembly responded to this by launching negotiations to formulate an International treaty on global climate protection, which resulted in completion of the United Nations Framework Convention on Climate Change (UNFCCC) in May 1992. The Convention on Climate Change sets an overall framework for inter-governmental efforts to tackle the challenge posed by climate change. It recognizes that the climate system is a shared resource whose stability can be affected by industrial and other emissions of carbon dioxide and other greenhouse gases. The Convention enjoys near universal membership with 189 countries having ratified it. The Convention entered into force on 21st March, 1994.

The Convention was opened for signature at the Earth Summit in Rio de Janeiro in June 1992, when it was signed by 154 states and European Community. It entered into force on March 21, 1994. India signed the UNFCCC on 10th June, 1992 and ratified it on 1st November, 1993. Under the UNFCCC, developing countries such as India do not have binding GHG mitigation commitments in recognition of their small contribution to the greenhouse problem as well as low financial and technical capacities. The Ministry of Environment and Forests is the nodal agency for climate change issues in India. The Convention established the Conference of Parties (CoP) as its supreme body. During CoP-3 meeting in Kyoto, Japan, the Parties agreed to a legally binding set of obligations for 38 industrialized countries and 11 countries in Central and Eastern Europe, to return their emission of GHGs to an average of approximately 5.2 per cent below their 1990 levels over the commitment period 2008-2012. This is called Kyoto Protocol to the convention. The Protocol entered into force on 16th February, 2005 and targets six main greenhouse gases: carbon dioxide (CO_2), methane (CH_4), nitrous oxide (N_2O), hydrofluorocarbons (HFCs), perfluorocarbons (PFCs) and sulphur hexafluoride (SF_6).

Article 12 of the Kyoto Protocol provides for the Clean Development Mechanism (CDM), which enables developing countries to participate in joint greenhouse gas (GHG) mitigation projects. Under this Protocol, Annex-I countries (developed countries and economies in transition) are required to reduce GHG emissions to below their 1990 levels.

The CDM enables these countries to meet their reduction commitments in a flexible and cost-effective manner. It allows public or private sector entities in Annex-I countries to invest in GHG mitigation projects in developing countries. In return, the investing parties receive credits or Certified Emission Reductions (CERs) which they can use to meet their targets under the Kyoto Protocol.

While investors profit from the CDM projects by obtaining reductions at costs lower than in their own countries, the gains to the developing country host parties are the form of finance, technology and sustainable development benefits.

The basic rules for the functioning of the CDM were agreed on at the Seventh Conference of Parties (CoP-7) to the UNFCCC held in Marrakesh, Morocco in October-November 2001.

Deemed as the "Kyoto Surprise", CDM is the only link between the Developed and developing countries under Kyoto Protocol (Jolly and Bajaj, 2007). Its workability will help ensure the effectiveness of the Kyoto Protocol and of developing countries' willingness to participate in a future global emissions regime. Since emissions from developing countries will eventually surpass those from the Annex-I countries, developing countries' full participation in such a regime is crucial.

India's Initiative

India and UNFCCC

India signed the United Nations Framework Convention on Climate Change on 10th June, 1992 and ratified on 1st November, 1992. India has undertaken numerous response measures that are contributing to the objectives of the United Nations Framework Convention on Climate Change. They are increasing significantly the capacity of renewable energy installations, improving the air quality in major cities and enhancing afforestation, thus putting economic development on a climate-friendly path.

Under the framework Convention, all parties are obliged to develop and publish a national inventory of Greenhouse Gases (GHG) not controlled by the Montreal Protocol. Articles 4(1) and 12(1) of the Convention provides for the parties to develop and publish a national inventory of GHG emissions and removal by sinks of all greenhouse gases not controlled by the Montreal Protocol.

There is a difference in fulfilment of this requirement between Annex countries and non-Annex-I countries. Annex-I Parties include the industrialized countries that were members of the OECD (Organisation for Economic Co-operation and Development) in 1992, plus countries with Economies In Transition (the EIT parties) including the Russian Federation, the Baltic States, and several Central and Eastern European States. Non-Annex-I parties are mostly developing countries. The former are required to report annual inventories, while the latter are required to report for the year 1994, or alternatively from 1990 from initial national communications and for thc ycar 2000 in the second national communications.

India and Kyoto Protocol

By acceding to the Kyoto Protocol on 26th August, 2002, "India has sent a good signal by taking the lead in the region and showing that multilateral approach is better than unilateral".

Right now India is emerging as one of the major beneficiaries of clean technology. India has necessary institutional and regulatory mechanism for implementation of CDM projects and has large potential for CDM related project activities in the areas of energy, coal, industry, renewable, transport and municipal solid waste.

Conclusion and Suggestions

In fact, environmental protection is a big challenge, which has been taken by the world community, through general and special standards, both at the national as well as international levels. However, as indicated by GEO-3 instead of being solved the problem in increasing day by day. The UNEP Global Environment Outlook (GEO) project was initiated in response to the environmental reporting requirements of Agenda 21 and to a UNEP Governing Council decision of May 1995, which requested the production of a comprehensive global state of the environment report. The recent GEO 3 has projected an outlook for 2002-3 on the basis of our scenarios namely, market, policy, security, and sustainability. It has emphasized that the next 30 years will be as crucial as the past 30 for shaping the future of the environment.

Old troubles will persist and fresh challenges will emerge, as upon our fragile resources increasingly heavy demands are being placed. It appears that the present approach of the world community to solve the problem is superficial. It does not offer a permanent solution due to certain inherent drawbacks such as: the function of international law is to promote harmonious and creative world order. It means the international environmental law has established only standards of behaviour, which differ widely in their nature, scope and implementation in different geographical and political regions. Deficiencies in the design and implementation of environmental laws constitute a major hurdle to environmental protection and sound natural resource management (Desai, 2002).

REFERENCES

Adam, David. 2007. "Move to cut the methane emissions by changing cows diet." *The Guardian*. July 10th Issue.

Aldy, J.L., Corvalan, C.F. and Philippeaux, H. 2003. Climate Variability and Change and their Health Effects in the Caribbean: Information for Adaptation Planning in the Health Sector. Conference 21-22 May, World Health Organization, Geneva.

A.P. Pollution Board v. *Prof. MV Nayudu*. 1999. 2 SCC 718 at 733-34.

Barrett, S. 2003. *Environment and Statecraft*. Oxford University Press, New York.

Buesseler, K.O. and P.W. Boyd, 2003. "Will Ocean Fertilization Work"? *Science*, 300:67-68.

Desai, Justice Ashok. 2002. *Environmental Jurisprudence*. Modern Law House, Allahabad, 2nd ed.

Falcam, L. 2001. "Death by Warming." *Honolulu Advertiser*. August 12.

Grabherr, G.; M. Gottfried and H. Pauli, 1994. "Climate effects on mountain plants." *Nature*, 369:448.

Hambro, E. 1994. "The Human Environment—Stockholm and After". *Year Book on World Affairs*, 20 (1974).

http://news.bbc.co.uk/2/hi/uk_news/wales/mid_/6288012.stm

http://nzsap.org.nz/proc/2004/ab04039.html

http://sustainabletable.org/introdictionary/

http://tracer.env.uea.ac.uk/soiree/

http://www.agr.state.nc.us/cyber/kidsworld/plant/nutrient.htm

http://www.croptrust.org/main/donors.php.

http://www.eatwild.com/cla.html

http://www.epa.gov/agriculture/ag101/dairyglossary.html

http://www.niwa.cri.nz/pubs/wa/10-4/newsforum

http://www.planktos.com/main.htm

http://www.timesonline.co.uk/tol/news/uk/science/article2051364.ece

Jacob Werksman. 1996. *Greening International Institutions*. Earthscan Publications Ltd., London.

Jacoby, Henry D., and A. Denny Ellerman, 2004. "The Safety Valve and Climate Policy." *Energy Policy*, 32 (4); 481-491.

Mendelsohn, Robert. 2003. "Assessing the market damages from climate change." In *Global Climate Change: The Science, Economics, and Politics*. Cheltenham, UK.

Olso, J.M. 1990. *Shifting the Burden of Proof, 20 Environmental Law at 898* (1990) quoted in A.P. Pollution Control Board Case (1992), 2 SCC at 734-35.

Roan, Sharon. 1990. *Ozone Crisis, the 15 Year Evolution of a Sudden Global Emergency*. Wiley.

Santilli, M., *et al.* 2005. "Tropical deforestation and the Kyoto Protocol: An editorial essay", *Clim. Change*, 71: 267-276.

Stellina Jolly and Varun Bajaj. 2007. "Clean Development Mechanism; International Legal Systems Response to Global Warming." Conference Papers, Thirty-Sixth Annual Conference, *The Indian Society of International Law*, 24-25 March, 2007, pp. 223-244, p. 227.

Vellore Citizen's Welfare Forum v. *Union of India*. 1995. 5 SCC 647 at 658.

Weatherhead, E.C. and S.B. Andersen, 2006. "The search for signs of recovery of the ozone layer." *Nature*, 441: 39-45.

WHO. 2003. *Synthesis Workshop on Climate Change and Health*. 1-4 December, Republic of the Maldives. World Health Organization, Geneva.

3
Effect of Environment on Social Change

Aparna Rajput and *Shiwani Bhatnagar*

Introduction

Social Change ......a general term, which refers to: Change in the nature, the social institutions, the social behaviour or the social relations of the society, community of the people or other social structures or any event or action which affects the group of individuals who shared values and characteristics together.

Humanity is about to face the most devastating catastrophe that mankind has encountered since the Bubonic Plague during the Middle Ages. This largest threat to humanity is global warming.

Eventually, global warming will melt enough of the polar ice to modify the world's oceans in two ways. First, the oceans will start to desalinate, in other words the salt in the sea water will be more diluted by the fresh water from the ice. Second, the oceans' water temperature will drop because of the cold melting ice, and this will cause the oceans' currents to change. Since the world's ecosystem is regulated by the ocean's currents, this will cause another ice age. That's right, global warming will start another ice age. No one knows when this ice age will start, but the world's agriculture cannot sustain the world's population during an ice age. This may even cause the extinction of mankind. No one knows when this ice age will happen, but it is coming. We know that the cycle of temperature increases are followed by ice ages, but this time the changes are not natural. These changes are caused by manmade greenhouse gases released in the environment.

Therefore, scientists believe that the next ice age will start much sooner than the scientific evidence has proven to have happened in the past.

Let us review for a moment. Global warming is the effects of greenhouse causing gases released into the environment that increases the average temperature of the world, thus adding more energy into the atmosphere. This increase in temperature and energy will cause an increase in rain and hurricanes, and wreck havoc to the agriculture of the world's food supply. Eventually, enough of the world's icecaps will melt to increase the world's sea levels, thus further destroying the world's agricultural land. Furthermore, the melting ice water will disrupt the world's oceans that regulate the world's temperatures and cause an ice age. And finally, human civilization currently cannot survive an ice age.

Environment Turning Bad Day by Day

In every environmental sector, conditions have either failed to improve, or they are worsening:

Public Health

Unclean water, along with poor sanitation, kills over 12 million people each year, most in developing countries. Air pollution kills nearly 3 million more. Heavy metals and other contaminants also cause widespread health problems.

Food Supply

Will there be enough food to go around? In 64 of 105 developing countries studied by the UN Food and Agriculture Organization, the population has been growing faster than food supplies. Population pressures have degraded some 2 billion hectares of arable land—an area of the size of Canada and the U.S.

Freshwater

The supply of freshwater is finite, but demand is soaring as population grows and use per capita rises. By 2025, when world population is projected to be 8 billion, 48 countries containing 3 billion people will face shortages.

Coastlines and Oceans

Half of all coastal ecosystems are under pressure of high population densities and urban development. A tide of pollution is rising in the world's seas. Ocean fisheries are being overexploited, and fish catches are down.

Forests

Nearly half of the world's original forest cover has been lost, and each year another 16 million hectares are cut, bulldozed, or burned. Forests provide over US$400 billion to the world economy annually and are vital to maintaining healthy ecosystems. Yet, current demand for forest products may exceed the limit of sustainable consumption by 25 per cent.

Biodiversity

The earth's biological diversity is crucial to the continued vitality of agriculture and medicine and perhaps even to life on earth itself. Yet human activities are pushing many thousands of plant and animal species into extinction. Two of every three species is estimated to be in decline.

Global Climate Change

The earth's surface is warming due to greenhouse gas emissions, largely from burning fossil fuels. If the global temperature rises as projected, sea levels would rise by several metres, causing widespread flooding. Global warming also could cause droughts and disrupt agriculture. How people preserve or abuse the environment could largely determine whether living standards improve or deteriorate. Growing human numbers, urban expansion, and resource exploitation do not bode well for the future. Without practicing sustainable development, humanity faces a deteriorating environment and may even invite ecological disaster.

Taking Action

Many steps toward sustainability can be taken today. These include:

using energy more efficiently, managing cities better, phasing out subsidies that encourage waste, etc.

Stabilizing Population

While population growth has slowed, the absolute number of people continues to increase by about 1 billion every 13 years. Slowing population growth would help improve living standards and would buy time to protect natural resources. In the long run, to sustain higher living standards, world population size must stabilize.

Conclusion

So we the human beings are suffering from various negative effects of imbalances in environment but unfortunately we ourselves are responsible for this present situation. Like the burning of coal, oil, and natural gas, as well as deforestation and various agricultural and industrial practices, are altering the composition of the atmosphere and contributing to climate change. These human activities have led to increased atmospheric concentrations of a number of greenhouse gases, including carbon dioxide, methane, nitrous oxide, chlorofluorocarbons, and ozone in the lower part of the atmosphere. If every country made a commitment to population stabilization and resource conservation, the world would be better able to meet the challenges of sustainable development. Practicing sustainable development requires a combination of wise public investment, effective natural resource management, cleaner agricultural and industrial technologies, less pollution, and slower population growth.

REFERENCES

Hinrichsen, D. and Robey, Bryant. 2000. *Population and the Environment: The Global Challenge*. Excerpts from a Johns Hopkins University report.

Moore, Thomas Gale. 1996. *Health and Amenity Effects of Global Warming*, Hoover Institution, Stanford University. JEL Code: J17, J31, Q25.

Solomon, S.; D. Qin; M. Manning; Z. Chen; M. Marquis; K.B. Averyt; M. Tignor; and H.L. Miller (eds.) 2007. Extracted from Chapter 2 of

"IPCC, 2007: Climate Change 2007: The Physical Science Basis. Contribution of Working Group I to the Fourth Assessment Report of the Intergovernmental Panel on Climate Change. Cambridge University Press, Cambridge, United Kingdom and New York, NY, USA".

UNEP. 1997. *Common Questions about Climate Change*. United Nations Environment Programme—World Meteorological Organization.

4

The Issue of Global Warming : Politicized Consensus and Emerging Concerns

Shreena Keswani, Deepali Joshi and *Anjali Adhikari*

Introduction

After a long spell of total denial of global warming issues by developed nations, now, the world has accepted the severe impact of this matter. Earlier, these concerns were declared as fuss or exaggerated facts by parties with vested interest. The concerned activists were raising the evidences of climate change referring to an increase in average global temperatures with a finger towards human activities contributing to it. It was believed to be primarily due to increases in "greenhouse" gases such as Carbon Dioxide (CO_2). But these concerns were largely rejected by a section of world media and powerful politicians as unauthentic populist approach of sensitization. The world-powers, who stretched their denial for long until it became inevitable for them to accept it, were those countries that were evidentially responsible for this grave situation in front of the universe.

This chapter is an effort to analyze the seriousness and concerns of the world towards global warming in the light of the world summits and their outcomes held since the inception of this century.

What is the Greenhouse Effect?

- Energy from the sun drives the earth's weather and climate, and heats the earth's surface;

- In turn, the earth radiates energy back into space;
- Some atmospheric gases (water vapour, carbon dioxide, and other gases) trap some of the outgoing energy, retaining heat somewhat like the glass panels of a greenhouse. These gases are, therefore, known as greenhouse gases; and
- The greenhouse effect is the rise in temperature on Earth.

Six main greenhouse gases are carbon dioxide (CO_2), methane (CH_4) is 72 times as potent a greenhouse gas as carbon dioxide) and nitrous oxide (N_2O), plus three fluorinated industrial gases: hydrofluorocarbons (HFCs), perfluorocarbons (PFCs) and sulphur hexafluoride (SF_6). Water vapour is also considered a greenhouse gas.

Differences in Greenhouse Gas Emission around the World

As the World Resources Institute highlights there is a huge contrast between developed/industrialized nations and poorer developing countries in greenhouse emissions, as well as the reasons for those emissions. For example:

- In terms of historical emissions, industrialized countries account for roughly 80 per cent of the carbon dioxide build up in the atmosphere to date. Since 1950, the U.S. has emitted a cumulative total of roughly 50.7 billion tonnes of carbon, while China (4.6 times more populous) and India (3.5 times more populous) have emitted only 15.7 and 4.2 billion tonnes respectively (although their numbers will rise).
- Annually, more than 60 per cent of global industrial carbon dioxide emissions originate in industrialized countries, where only about 20 per cent of the world's population resides.
- Much of the growth in emissions in developing countries results from the provision of basic human needs for growing populations, while emissions in industrialized countries contribute to growth in a standard of living that is already far above that of the average person worldwide. This is exemplified by the large contrasts in per capita carbon emissions between industrialized and developing countries. Per capita emissions of carbon in the U.S. are over 20 times higher than India, 12 times higher than Brazil and seven times higher than China.

At the 1997 Kyoto Conference, industrialized countries were committed to an overall reduction of emissions of greenhouse gases to 5.2 per cent below 1990 levels for the period 2008-2012.

Who is mainly Responsible for the Turmoil

The UNFCCC reported (November 17, 2008) that although industrialized nations have reduced emissions between 1990 and 2006, in recent years, between 2000 and 2006, greenhouse gas emissions have generally increased by 2.3 per cent. The previously 15 members of European Union are large emitter. They account for 10 per cent of global emission and 24 per cent of industrialized countries manmade emission of six main gases.

Some nations with large reductions are also seeing limits, for example:

- UK (15.1% reduction) benefited by switching from coal to natural gas but that switch is largely in place now.
- Germany (18.2% reduction) has certainly invested in greenhouse gas emission reductions, but has been helped in large part because of reunification.

It has been known for some time now that developing countries including India will be affected the most. Reasons vary from lacking resources to cope, compared to developed nations, immense poverty, regions that many developing countries happen to be the ones where severe weather will hit the most, small island nations area already seeing sea level rising, and so on.

World View on Global Warming : Proceedings of the Summits

Keeping in view the harmfulness of greenhouse gases and their long-term effect on developing countries, there was an urgent need to look into this issue. United Nations and other world organizations took lead and conferences were initiated at world level in different parts of world involving different nations to reach on a common consensus for a proper remedy for global warming. The discussions held in these summits reveal the attitude of the developed nations towards this grave concern. It also depicts the pressure and its consequences on the shifting attitude of leading nation.

UN Conference on Environment and Development—The Earth Summit

This conference was held in Rio de Janeiro (Brazil) in 13-14 June, 1992. About 172 countries participated in this conference.

Principal Themes

- Environment and sustainable development.
- The summit message was nothing less than a transformation in our attitudes and behaviour that would bring about necessary changes was transmitted.

Outcomes

- Production of toxic components, such as lead in gasoline or poisonous waste, is being scrutinized in a systematic manner by UN Government.
- Alternative sources of energy are being sought to replace the use of fossil fuels which are linked to global climate change.
- Now check on public transportation system is being emphasized in order to reduce vehicle emission, congestion in cities and to avoid health related problems.

Kyoto Protocol Climate Conference

The Kyoto Protocol was the climate change treaty negotiated in 1997, setting targets for emissions of greenhouse gases. The Conference of Parties III (CoP-3), Kyoto, Japan, the Kyoto Conference on climate change took place. There, developed countries agreed to specific targets for cutting their emissions of greenhouse gases. A general framework was defined for this, with specifics to be detailed over the next few years. This became known as the Kyoto Protocol.

Outcomes

- The US proposed to just stabilize emissions and not cut them at all, while the European Union called for a 15 per cent cut. In the end, there was a trade off, and industrialized countries were committed to an overall reduction of emissions of greenhouse gases to 5.2 per cent below 1990 levels for the period 2008-2012.

- A lot of primarily industry arguments against the Kyoto Conference and Global Warming in general, claim that it will hurt the global (or USA's) economy and affect people's jobs. Some of the well-respected scientists claiming that Global Warming is a myth have been sponsored in some way by various commercial interests as well.
- At the end of March 2001, U.S. President George Bush said that he "opposed the Kyoto Protocol." One of the reasons he cited was because India and China would not be subject to Kyoto measures and would increase their emissions. Yet he ignored that on a per capita basis, India and China's emissions are far less than the United States.
- Economic fears were also cited. Business interests. Bush's announcement resulted in almost world-wide condemnation and many pointed to Bush's controversial election which included backing and contribution of millions of dollars by the oil industry, as factors in this decision. In February 2002, Bush has proposed a different means to address Climate Change instead of following the Kyoto plans, which again met with the same criticism. An interesting thing to note has been his Administration's play on words, saying greenhouse gas intensity will be reduced, rather than emissions.

Professor of economics, Paul Krugman from M.I.T., points out the issue in *New York Times*: "What is this thing called greenhouse gas intensity? It is the volume of greenhouse gas emissions divided by gross domestic product. The administration (of George Bush) says that it will reduce this ratio by 18 per cent over the next decade. But since most forecasts call for G.D.P. to expand 30 per cent or more over the same period, this is actually a proposal to allow a substantial increase in emissions."

- Also, the US wishes to suggest an emission trading policy instead of the suggested policies of the Kyoto Protocol. However, it has been criticized as an ineffective and inequitable system.

Buenos Aires Climate Conference

So, following the Kyoto Protocol, the next round of talks (November 2—November 13, 1998) were held in Buenos Aires, Argentina for the Fourth Meeting of the Conference of the Parties (CoP-4) to the UN Framework Convention on Climate Change (UNFCCC).

There are many issues that still need to be discussed, especially on the trading of Carbon emissions and equity between the rich and developing nations. The politics and self-interest that was at work at the Kyoto Conference was feared to manifest itself this time round. Even though the talks have now ended with some hailing it as a success, others criticize it for not having clear-cut deadlines, etc.

The Hague Climate Conference

November 13 to November 24, 2000 saw the sixth session of the United Nations Framework Convention on Climate Change (UNFCCC) Conference of the Parties. This one was intended to wrap up three years of negotiations on the implementation of the Kyoto Protocol.

Outcomes

- This has even caused some rifts between US and European climate and environment groups. US wanted carbon sinks as part of agreement (as well as other provisions like nuclear energy, etc.), which received enormous criticism. In addition, the Europeans were not swaying on their stance that the US should not be given exemptions and should not be allowed to meet much of its greenhouse gas reduction targets without actually cutting emissions.
- The developing countries were sidelined. Even the trade-offs being considered above were largely without considering the G77, or the developing Southern countries.

"The US, with 4 per cent of the world population, is responsible for 24 per cent of all carbon emissions. And while the dirtiest nation swaggers around the globe policing how the rest of the world trades and organises its finances, it apparently believes it should come under no scrutiny at all for running its own industry with fuel-inefficient, environmentally damaging technologies long abandoned by Europe—Dirty Uncle Sam wrecks deal on global warming."

—*The Guardian*, November 26, 2000

Marrakesh Climate Conference

From 29th October to 9th November, 2001, governments met in Marrakesh, Morocco, for the 7th Conference of the Parties to the UN Framework Convention on Climate Change (CoP-7). The purpose of the meeting was to agree legal text covering outstanding technical aspects of the political agreement reached in Bonn in July 2001 on how to implement the Kyoto Protocol. U.S. was not there as they had pulled out of talks earlier in 2001.

This meeting ended with agreement on how to enforce the Kyoto accord on tackling climate change. As the previous link reports, there were some further concessions to be made due to "co-ordinated obstruction" by nations such as Japan, Canada and Russia that wanted things like carbon sinks to be accounted.

Delhi Climate Conference

October 23 to November 1, 2002 saw some 180 countries converging in New Delhi for the Eighth Conference of Parties (CoP-8) to the United Nations Framework Convention on Climate Change (UNFCCC). Government delegates, representatives from inter-governmental and non-governmental organizations, media and business—all attended the conference. The meeting's aim was to see the formulation of the "Delhi Declaration".

Outcomes

- Rich countries turn the debate around to poor countries. Throughout this conference developed countries kept up intense pressure on developing countries' commitments through repeated insinuations in speeches and statements.

Buenos Aires Climate Conference

The tenth session of the Conference of the Parties to the U.N. Framework Convention on Climate Change (CoP-10) took place in Buenos Aires in mid-December 2004.

Discussions at CoP-10 highlighted a range of climate-related issues including:

- The impacts of climate change and adaptation measures.
- Mitigation policies and their impacts.
- Technology.
- Entry into force of the Kyoto Protocol.

Impacts and Outcomes

- Social Justice and Equity Concerns were ignored. When the United Nations Framework Convention on Climate Change was formulated and then signed and ratified in 1992. The principle of common but differentiated responsibilities was acknowledged. In short this principle recognized that:
  1. Today's rich nations are the ones responsible for global warming as greenhouse gases tend to remain in the atmosphere for many decades, and rich countries have been industrializing and emitting climate changing pollution for many more centuries than the poor countries.
  2. It is, therefore, unfair to expect the Third World to make emissions reductions and also unfair considering their development and consumption is for basics and for developing, while for the rich, it has moved on to luxury consumption and lifestyles.
- Furthermore, developing countries too were to reduce emissions ultimately, but in a different way: The rich were to help provide means for the developing world to transition to cleaner technologies while developing.

The CoP-10 meeting had ended with an agreement to hold a seminar of governmental experts in 2005 to "exchange information" about ways to fight global climate change.

Discussions of post-Kyoto

At this meeting, there were attempts to discuss beyond 2012 which is when the Kyoto Protocol should be implemented by. In these "post-Kyoto" discussions, developing nations feared that they would be forced into reduction commitments, so some developing countries attempted to resist discussing this.

Montreal Climate Conference

The two-week climate change conference was held in Montreal, Canada, at the beginning of December 2005. This involves practically the whole world—189 countries—that are part of the United Nations Framework Convention on Climate Change (UNFCCC). A number of issues were to be discussed as a result, including :

- Clarifying and agreeing the rules and commitments that different countries have;
- How to help developing countries reduce their emissions;
- How to measure emission reductions;
- Accountability.

Outcomes

- The Kyoto Protocol entered into force from 16th February, 2005 with more than 30 industrialized countries bound by specific and legally binding emission reduction targets. As a first step, these cover the period 2008-2012. This meeting confirmed a number of implementation plans, such as emissions trading, joint implementation, and clean development mechanisms. These last two are supposed to help developing countries while recognizing the industrialized countries' contributions. It was agreed that sustained discussions with all countries was needed on various things such as:
    1. Advancing development goals in a sustainable way for poorer countries;
    2. Addressing action on adaptation to climate changes;
    3. Pursuing technology development and technology transfer to help develop solutions; and
    4. Exploring further the potential of market-based opportunities.

- The Montreal meeting brought some attention to developing country commitments and some rich countries have also been raising that issue (some say also to deflect some criticism of themselves for not reducing as much emissions as they should be, or for even increasing them). In particular, various nations say that China (the second largest emitter of carbon dioxide) and India are rapidly growing and so their emissions are going to increase as well, even overtake some industrialized countries in the next few years. Yet, the Kyoto Protocol does not demand targets from these countries. Hence, nations such as the US, and even recently, a few European nations, say this is unfair.

- China and India in particular, however, are industrializing. As their role and impact on the planet changes, they will be expected also to change their habits in the same way today's industrialized nations are expected to. So it is fair to expect China and India to be involved in the near future, but perhaps unfair to be involved immediately, or at least to the same effect as industrialized nations immediately. For most of the other poorer countries, as explained above, it would be an oversimplification to expect them to also have reduction targets to the same extent as rich countries, until they too start to industrialize.

Yet, these political issues, as crucial as they are, also detract from the other major problem: the changing climate is not going to wait for anyone, no matter how fair or unfair it seems. What environmentalists and others, therefore, feel that Montreal has at least given is a new impetus to get moving on future discussions.

Bali Climate Conference

The UN Conference on climate change held in Bali, Indonesia in December 2007 led to a final agreement known as the "Bali Roadmap". The Bali Roadmap outlined a new negotiating process to be concluded by 2009 to feed into a post-Kyoto (i.e. a post-2012) international agreement on climate change.

The Roadmap included a decision to launch an Adaptation Fund as well as further decisions on technology transfer and on reducing emissions from deforestation.

But the conference was also accompanied by controversy, including :

- The US position being at odds with most of the rest of the world.
- Talk of developing countries' responsibilities (such as China and India) while rich countries (the source of the problem) have made little progress, themselves.

Poznań Climate Conference

Poznań, Poland was the venue for the 14th United Nations Climate Change Conference, also known as CoP-14, between December 1 and December 12, 2008.

Outcomes

As summarized by the BBC, the main outcomes from the Poznań seem to be the following :

- Management of a UN Adaptation Fund to help developing countries (who will worst hit by the effects of climate change) agreed.
- Funds can now be disbursed using a 2 per cent levy on carbon trading under the UN Clean Development Mechanism.
- Progress on how environment-friendly technology can be transferred to developing countries.
- Agreement that deforestation needs to be reduced.
- Recognition that the situation is quite urgent.

Parallel EU Meeting in Brussels

A two-day summit in Brussels, ended on the same day as the Poznań Climate Conference ended. The EU summit outcome was a bit surprising given EU's generally positive step to addressing climate change in the past. As summarized by Inter Press Service, the summit saw a drastic amendment to the way energy-intensive firms pay for their emissions of greenhouse gases.

Some groups claimed the EU had watered down the initial proposals. Germany, for example, was accused of wanting to protect its large manufacturing sector during its recession, while Poland and Hungary for example, and depend a lot on coal.

The EU did agree, however, to a 20 per cent cut in greenhouse gas emissions by 2020, compared with 1990 level. But this too seems less than ideal as the Inter-governmental Panel on Climate Change has said that 25-40 per cent reduction is needed by then.

This all comes soon after the UNFCCC reported (November 17, 2008) that although industrialized nations have reduced emissions between 1990 and 2006, in recent years, between 2000 and 2006, greenhouse gas emissions have generally increased by 2.3 per cent.

This is despite an overall decrease of 4.7 per cent since 1990. However, the more recent period suggests the rich country emission reductions are not sustainable. Furthermore, it looks worse considering a large part of this decrease is because of the collapse of the Soviet Union. As transition economies started to recover around 2000, emissions have started to rise.

Copenhagen Climate Conference

December 7—December 18, 2009, Copenhagen, Denmark was the venue for the 15th annual United Nations Climate Change Conference, also known as the 15th Conference of the Parties or CoP-15. Instead of a positive outcome, most commentators saw it as a failure, though for different reasons.

As the Copenhagen Conference drew closer, it was becoming more likely that nothing concrete would be achieved due to various impasses.

Four essential issues were decided to address in Copenhagen Conference :

1. How much are the industrialized countries willing to reduce their emissions of greenhouse gases?
2. How much are major developing countries such as China and India willing to do to limit the growth of their emissions?
3. How is the help needed by developing countries to engage in reducing their emissions and adapting to the impacts of climate change going to be financed?
4. How is that money going to be managed?

Despite the pressure and expectations, including gruelling last minute efforts to salvage something, all that emerged was a vague agreement—the Copenhagen Accord.

Key points in accord included :

- To keep the maximum temperature rise to below 2°C.
- To list developed country emission reduction targets and mitigation action by developing countries for 2020.
- $30 billion short-term funding for immediate action till 2012.
- $100 billion annually by 2020 in long-term financing.
- Reiterating past intentions such as providing mechanisms to support technology transfer and forestry.

While many were disappointed with the final outcome, many (especially some politicians at the Conference who perhaps had to avoid losing face) tried to put a positive spin on things.

Common but Differentiated Responsibility Principle Sidelined Again

As Inter Press Service (IPS) summarized: What is abundantly clear is the enormous divide between the rich and poor countries. Poor countries want deep cuts in emissions by the industrialized world, and the latter continue to resist significant cuts and legally binding targets.

Year after year at climate summits, it seems this principle is often ignored by some rich nations and their media. The US and others have characterized the campaign for climate justice and equality to the atmosphere as a way to claim climate “reparations”; that it is unfair to make the industrialized nations pay for climate emissions into the past century or more at a time when they did not know it would cause more harm. That seems reasonable. However, one of the implications of any agreement that is subsequently drawn up will, in effect, put disproportionately more burden on the poorer countries to tackle a problem they did not largely cause. The poor are less likely to have the resources to do so, which also means that tackling climate change is less likely to be successful.

Nonetheless, actions from China at the Copenhagen summit were a cause for concern. While seemingly backed into a diplomatic corner by the US and others on the issue of transparency and emission reduction targets, it flexed its growing muscle, possibly a bit too far, showing the world that it can be just like the superpowers that have come before it and disregard many opinions and views as it sees fit.

The meeting was also interesting for the geopolitical implications of China's actions: it almost certainly highlighted Europe's decreasing influence even as a united bloc.

Obstacles and Slow Progress

Many of the objectives highlighted above have still not been recognized. For example, the industrialized countries have not provided much help in many areas such as effective emission reductions and stalling on developing country commitments, or opposing the Kyoto Protocol itself.

Greenhouse Gases and Emissions resulting from Human Activity

Every few years, leading climate scientists at the UN's Inter-governmental Panel on Climate Change (IPCC) have released major, definitive reports detailing the progress in understanding climate change. From the outset they have recommended that there be emission reductions. This body is comprised of hundreds of climate scientists around the world.

At the beginning of January 2007, the IPCC's fourth major report summarized that they were even more certain than the trends observed before. Global atmospheric concentrations of carbon dioxide, methane and nitrous oxide have increased markedly as a result of human activities since 1750 and now far exceed pre-industrial values determined from ice cores spanning many thousands of years. The global increases in carbon dioxide concentration are due primarily to fossil fuel use and land-use change, while those of methane and nitrous oxide are primarily due to agriculture.

The understanding of anthropogenic warming and cooling influences on climate has improved since the Third Assessment Report (TAR), leading to very high confidence that the globally averaged net effect of human activities since 1750 has been one of warming. Most of the observed increase in globally averaged temperatures since the mid-twentieth century is very likely due to the observed increase in anthropogenic greenhouse gas concentrations—an induced climate change because of better scientific understanding.

Can there be a Common Course of Action?

Despite all the optimism that these problems can be solved, the concerns have been raised by many for decades. Ultimately, it seems, we are showing our future generations that once again power, greed, selfishness and other negative qualities of human nature can easily usurp any positive traits such as cooperation. For climate negotiations, many now hope the follow-up of the meeting in Mexico held in 2010, which is likely to proceed towards concrete agreements. Many scientists say greenhouse emissions need to reach the peak sooner rather than later, so each year seems like wasted time. This big lurking challenge is still on head but we are not convinced to become united and unanimous for the action against global warming. Summit after summit, we are still a divided house. Challenge is aggravating moment by moment and we are still hanging to see the doom's day.

5

Exploring the Aspects of Women in the Changing Climate

Hema Bhatt and *Savita Rathour*

Introduction

Ever since the earth came into being there has been a climate system. The climate of a place is the average weather that it experiences over a period of time. The factors that determine the climate at a location are the rainfall, sunshine, wind, humidity, and temperature. While changes in the weather may occur suddenly and noticeably, changes in the climate take a long time to settle in and are, therefore, less obvious. Throughout the earth's history there have been changes in the climate. There have been well-marked cold and hot periods and all life forms adapted naturally to this change. Over the last 150-200 years climate change has been taking place too rapidly and certain plant and animal species have found it hard to adapt. Human activities are said to be responsible for the speed at which this change has occurred and it is now a cause of worry to scientists. The drastic change climate now is not only the scientific issue; it requires contribution at individual and community level.

Women around the world play a key role in the protection of biological diversity through their various responsibilities. They have recognized the need not only to protect the biodiversity, but also to reshine and recreate it. Wangari Maathai founded Kenya's green belt movement, involving 80,000 women in planting of trees. Women are a largely untapped resource that society must use to effectively and justly combat climate change. They need to be harnessed to prepare communities for global warming's effects, particularly in developing countries where warming will have the most severe consequences.

According to a new report, global warming is not gender-blind either; women are especially vulnerable to its effects, making up a shocking 80 per cent of climate refugees. According to a briefing released by the U.K.-based Women's Environmental Network (WEN), more than 10,000 women die each year from weather-related disasters such as tropical storms and droughts, compared to about 4,500 men. Women also comprise 20 million of the 26 million people estimated to have been displaced by climate change. These discrepancies are especially strong in countries that lack gender equality.

For example, in the Bangladesh cyclone in 1991, almost five times more young women died than men. This was largely because women could not swim, had restricted mobility because of their clothing, and many women left their homes too late because they waited for a male relative to accompany them. Men were able to warn each other of the danger as they met in public spaces, but did not always get the information back to their families. A study of disasters in 141 countries found that, where women had equal rights, there was little or no difference in the number of women and men that died, but where women's rights were compromised, female mortality was higher.

Role of Men and Women

Because men and women are likely to have different roles in the family, community and workforce, they are likely to have different personal priorities when it comes to environmental protection at the local and community level. Men and women are also likely to have different interactions with the environment, which present each gender with different opportunities to protect it. Moreover, it is likely that men and women adopt different strategies and are the sources of different types of knowledge regarding environmental protection. For women with double and triple roles as breadwinner, homemaker and community manager, time pressures may force them to adopt unsound environmental practices at the household level, e.g., in disposing of waste and recycling. Access to other "time saving" appliances may be viewed as a top priority for women, even though these may not make efficient use of natural resources.

Engaging Women on Adaptation Strategies

Women in any country have an intimate knowledge of the social and natural systems which is affected by global warming or which can affect the global environment. They are at the heart of their families' and communities' resource management and well-being. Women are more likely to be hit harder by climate change than men as they have more direct contacts with environmental issues due to their social and family roles and the simple fact that a majority—as much as 70 per cent—of the world's poor are women. One researcher concludes that women are 14 times more likely to die than men in a natural disaster such as a *tsunami*. But over the last few years the increasing portfolio of climate solutions is beginning to include gender-sensitive approaches and women's involvement. Observers realize that women need to be protected, engaged, and empowered for climate solutions to truly succeed. They also see that a vulnerable segment of the population is in fact one with mass potential to bring positive change. Women and girls, for example, are responsible for gathering cook stove fuel and producing 60 to 80 per cent of the food in developing countries. According to the United Nations, women in Sub-Saharan Africa spend 40 billion hours a year collecting water. They can, therefore, provide valuable insight when it comes to formulating adaptation policies and implementing projects as fuel and water become scarcer and agricultural yields shrink.

U.N. officials, UNFCCC country delegates, and non-governmental organizations are beginning to integrate gender into their discourse and policy pushes. A handful of NGOs have more publicly championed ìn gender justice, such as the international anti-poverty organization CARE and the Women's Environment and Development Organization, or WEDO. They held side events at the Copenhagen climate change conference last year on this very issue, and groups like Climate Wise Women have since taken their message on the road.

Groups like WEDO continue to push the United Nations to involve women at high-level climate negotiations. They note that at Copenhagen 30 per cent of the country delegates were women. On the one hand, this was the peak for women's engagement, but on the other hand, it does not mirror reality. Similarly, in March, U.N. Secretary General Ban Ki-moon put together the Advisory Board on Climate Change Financing and appointed all men. The gender justice community subsequently voiced their concern over the apparent apathy to involve women in high-level discussions.

The first step to making women full participants in adaptation is recognizing their value in local climate negotiations and supporting advocacy group work. Next is finding concrete ways to integrate women into the planning, development, and execution of climate adaptation strategies. Climate adaptation strategy literature is minimal, but observers often turn to experience with disaster risk reduction to craft informed approaches. The disaster risk reduction community is also giving gender issues more consideration, which involves evaluating regional, national, and local response systems through a gender lens, collecting supplemental data, and ensuring women have equal access to early warning systems and the resources for preparedness and disaster assistance. Further, these techniques are made much more possible and effective with targeted financing, which is why some advocates argue that part of adaptation funding should be set aside for gender-specific training, community workshops, and disaster plans.

The Steps Women can Forward for Environment Conservation

Let us think here, how women can contribute to, how they can participate in, and how they can play their role in protection of the environment.

Population Control and Environment Conservation

As the human population continues to increase so does the magnitude of the burden placed on the Earth's resources. The links between population and environmental quality are complex and varied. Humans are depleting natural resources, degrading soil and water, and creating waste at an alarming rate. The challenge of preserving a sustainable environment is greatest in poor countries that already struggle to meet their people's needs and that face the greatest population growth. Worldwide population growth is a delicate climate change issue, but the impact it has on the environment cannot be ignored. Family planning and encouraging a voluntary choice for smaller families is a feasible solution.

Slowing the rate of population growth may give countries time to take measures to meet people's needs, while protecting the environment through various means. The search for fuel wood, water and other basic needs makes the poor unwitting agents of environmental change. In poor countries, most of these agents are women and children. Preventing unwanted pregnancies in developing countries through family planning might be one of the most cost-effective ways to preserve the environment. Here the one who can help in sustaining the issue is the most concerning member of family, the adult female member. Family planning is a sensitive issue and it can be tackled only by sensitizing the women. In poor and uneducated families, it is hard to convince them for accepting the population control measure. But a woman can understand the problems of a large family and the relative consequences of higher consumption and high demands of resources in the family. Woman can shoulder this responsibility very well.

Waste Management

Usually, people are in habit of throwing the rubbish and the waste after cutting the vegetables, graining, wiping and cleaning the floor, on the municipal road, just outside the houses. This creates nuisance which gives birth to mosquitoes and malaria or other viral fever. Hence everyone should use the garbage box. The collected garbage should be thrown in the garbage boxes kept by the municipality. If this habit is developed, women not only be able to keep their houses clean but, the colony, the city and their surrounding environment will also become orderly, tidy and healthy.

Gardening

We all are familiar with the fact that these days houses are very small. Our women can come forward by developing hobby of gardening. They can at least keep the environment of their houses healthy. Plants add to the beauty of the house, provide freshness to the eyes and are helpful in keeping the nearby environment clean. Questions may crop up in one's mind that "can this hobby be carried out in such times of exorbitant prices when it has become difficult to make both ends meet?" Answer to this question is in the affirmative. Why do we go farther away? *Tulsi* plant is being worshipped since long, is it very costly? It is a good source of oxygen as well as this plant has medicinal value. Are crotons very costly? Vegetables like Turai "Beans" do not need much space to grow. Besides keeping the surrounding atmosphere healthy, money is also saved. Saplings of plants can also be shared with neighbours. Again, our women can make use of their management ability by using empty cans, bottles, tins, etc., as pots for growing the plants. Because women often play the role of "carer" or "healer" in the community, they may possess knowledge of environmentally sustainable practices, such as benefits derived from medicinal plants and other non-timber forestry products.

Non-Conventional Sources of Energy

India has limited resources of conventional energy. Dependency of our mass population for fuel on forests is another additional cause for the vanishing forest wealth and disturbing ecosystems. Wood remains the main source of energy in many urban and rural areas of the developing countries. In developing countries, like India, women are collecting fuelwood from nearby forests, carrying it for domestic purposes and also selling it elsewhere to supplement their family incomes. The firewood, etc., burnt in open country *chulha* wastes 82 per cent of the fuel, since efficiency of heat utilization is only 12 per cent (UNICEF, 1991). Out of total amount of wood harvested in India, 42.3 per cent is burnt as firewood and charcoal. Soil, water and forest can be conserved if alternative kitchen fuel is provided. Women can again contribute in this direction and come forward by using solar lights, solar geyser, solar cooker, smokeless *chulhas*, etc.

Development of Good Habits among Children

Mother is the first teacher. It is she who plays an important role in the development of the personality of the child. She can instil in future generations a respect for nature and the value of biodiversity. She can develop good habits in the child from the very beginning as they should not allow the children to waste the paper—the children are usually in habit of tearing away papers from their notebooks, a mother can keep a check on this bad habit of the child. She can make them understand that for paper production, the wood is used and deforestation can bring unhappiness by disturbing the ecosystem.

Light Bulbs Matter

If every household in the United States replaced one regular light bulb with an energy-saving model, we could reduce global warming pollution by more than 90 billion pounds over the life of the bulbs; the same as taking 6.3 million cars off the road. So, replace your incandescent bulbs with more efficient compact fluorescents, which now come in all shapes and sizes. You will be doing your share to cut back on heat-trapping pollution and you will save money on your electric bills and light bulbs. Save electricity and reduce global warming by turning off lights when you leave a room, and using only as light as you need. And remember to turn off your television, video player, stereo and computer when you are not using them. It is also a good idea to turn off the water when you are not using it. While brushing your teeth, shampooing the dog or washing your car, turn off the water until you actually need it for rinsing. You will reduce your water bill and help to conserve a vital resource.

Reduce, Reuse, Recycle

Do your part to reduce waste by choosing reusable products instead of disposables. Buying products with minimal packaging (including the economy size when that makes sense for you) will help to reduce waste. And whenever you can, recycle paper, plastic, newspaper, glass and aluminium cans. If there is not a recycling programme at your workplace, school, or in your community, ask about starting one. By recycling half of your household wastes, you can save 2,400 pounds of carbon dioxide annually. Avoid products that come with excess packaging, especially molded plastic and other packaging that can not be recycled. If you reduce your household garbage by 10 per cent, you can save 1,200 pounds of carbon dioxide annually. Women's domestic responsibilities (e.g., procuring fuel and water) may mean that they adopt unsound environmental practices when these resources are scarce.

Plant a Tree

If you have the means to plant a tree, start digging. During photosynthesis, trees and other plants absorb carbon dioxide and give off oxygen. They are an integral part of the natural atmospheric exchange cycle here on Earth, but there are too few of them to fully counter the increases in carbon dioxide caused by automobile traffic, manufacturing and other human activities. A single tree will absorb approximately one ton of carbon dioxide during its lifetime.

Encourage Others to Conserve

Share information about recycling and energy conservation with your friends, neighbours and co-workers, and take opportunities to encourage public officials to establish programmes and policies that are good for the environment. Individual choices can have an impact on global climate change. Reducing your family's heat-trapping emissions does not mean forgoing modern conveniences; it means making smart choices and using energy-efficient products, which may require an additional investment upfront, but often pay you back in energy savings within a couple of years.

Look for Energy Star : Get a Report Card from Your Utility Company

Many utility companies provide free home energy audits to help consumers identify areas in their homes that may not be energy efficient. In addition, many utility companies offer rebate programmes to help pay for the cost of energy-efficient upgrades. When it comes time to replace appliances, look for the Energy Star label on new appliances (refrigerators, freezers, furnaces, air conditioners, and water heaters use the most energy). These items may cost a bit more initially, but the energy savings will pay back the extra investment within a couple of years. Household energy savings really can make a difference: If each household in the United States replaced its existing appliances with the most efficient models available, we would save $15 billion in energy costs and eliminate 175 million tonnes of heat-trapping gases.

Conclusion

The Indian women were always ahead in the matter of prevention of pollution and protection, preservation, conservation, promotion and enhancement of the environment. They have made greater sacrifices; have exhibited greater love than their male counterparts in the area of environment protection. The strong desire, devotion and dedication towards better environment made Indian women a crusader against environment pollution. Today the torchbearers are Maneka Gandhi, an environmentalist and politician, Medha Patekar, a social worker and environmentalist who are carrying on the crusade of environment protection. But the problem of present day environment protection cannot be solved with the involvement of a few selected groups of women. It requires a global involvement of the entire community of womenfolk. Today, there is need of participation of every woman of every class/every locality, and any woman should not underestimate her role.

U.S. Secretary of State Hillary Clinton—while not explicitly a U.S. climate negotiator—has unabashedly placed women as a strategic centrepiece in her foreign policy approach. She appointed Melanne Verveer as her ambassador at large for global women's issues and stated, "Without providing more rights and responsibilities for women, many of the goals we claim to pursue in our foreign policy are either unachievable or much harder to achieve".

We should similarly place women on the front lines as we craft solutions to abate global warming and adapt to it. They embody much knowledge, responsibility, and unrealized potential, and are, therefore, essential if we are to achieve any meaningful degree of success. By giving women a greater voice in climate-change decision-making, and taking strong action on gender equality, not only could disastrous climate change be avoided but progress made toward a more equitable society.

REFERENCES

Campbell, N.A.; J.B. Reece, *et al.* 2008. *Biology*, 8th edition. Benjamin/ Cummings Publishing Company, Inc.

Jepma, J.C.; Mohan Munasinghe. 1998. *Climate Change Policy: Facts, Issues and Analyses*. Cambridge University Press, UK.

Kumar, A.; D.R. Jana and A. Kumar, 2008. Environmental Management System (EMS) and Sustainable Development. Arab Institute in Science and Engineering. *ARISER*, Vol. 4, pp. 195-204.

Mathews, W.H. 1977. *Resource Materials for Environmental Management and Education*. Pub: the MIT Press, Cambridge, Massachusetts,

Palucci, B.; A. Hall and N. Axinn, 1997. *Family Decision Making: An Ecosystem Approach*. John Wiley & Sons Inc, New York.

Planet green.com/global-warming-hits-worlds-women-hardest.php.htm

Venketswaran, S. 1995. *Environment, Development and the Gender Gap*. Sage Publication, New Delhi.

6

Impact of Climate Change on Nutritional Status of Pregnant Women

Shalini and *B. Singh*

Introduction

Uttar Pradesh (U.P.) is a major contributor to the national foodgrain stock, owing to fertile regions of the Indo-Gangetic plain and partly due to irrigation measures such as the Ganga Canal and tube wells. In 1991-93 F & HS (Family and Household Survey) revealed that 73 per cent of the population in the State was engaged in agriculture and 46 per cent of the State income, was accounted for by agriculture. U.P. has related its pre-eminent position in the country as a food surplus State.

There are basically four regions in U.P., namely Western, Eastern, Central, and Ruhelkhand. Western U.P. is more advanced and prosperous in terms of agriculture as compared to other regions of the State. Hathras district is situated in western region which is also known as Brij Bhumi (http://bit.ly/5MOKb). It is located in the middle Doab of Ganga and Yamuna river, right between the districts of Agra, Mathura and Aligarh. Here the soil is very fertile due to the plains of Ganga and Yamuna and hence all types of crops during the *rabi*, *kharif* and *zaid* seasons are grown here. There is hardly any drought, and floods are almost never heard of. Labour gets employment throughout the year, either in the fields or in nearby factories. There are number of factories of splitting legumes, ginning mills (cotton mills) and other agro-based industries. Men do heavy works which require high energy and women get work as a domestic helper or in agro-based industries such as pickle factories, carpet industries, splitting seeds of watermelon and cucumber, packing of puffed corn and puffed rice, etc. Usually low income group get sufficient food from their earnings. To fulfil their domestic needs, they have ration cards and other facilities provided by government and NGOs.

However, in the months of July, August and September 2009 situation was different. Out of 71 districts, 20 districts of U.P. were declared drought hit, following scanty rainfall. These districts recorded less than 40 per cent of the normal rainfall and sowing was reduced to 50 per cent of the regular. In August 2009, drought brought U.P. paddy output to just 6 million tonnes, which was 13.1 million tonnes last year; maize production also dropped down from 1.15 to 0.9 million tonnes (http://bit.ly/cqgNmL).

Drought is a climate, normally characterized by deficient supply of moisture, resulting either from subnormal rainfall, erratic rainfall distribution, higher water need, water resource mismanagement or is a combination of all the above said factors. Occurrence of drought is a gradual process posing threat to food security, health of population and accelerates human morbidity and mortality. Drought leads to water and food shortage and has a long-term environmental, economic and health impact on the population. Reduced food intake results in malnutrition and deficiency of energy, protein and several other micronutrients, which in turn sets off the vicious malnutrition.

Consequences of Malnutrition

1. Anaemia is the most common nutritional deficiency disorder in the world. WHO has estimated that prevalence of anaemia in developing countries among pregnant women is 51 per cent and it is 65-75 per cent in India. Prevalence of anaemia over all groups is relatively higher in India, in comparison to other developing countries. India contributes about 80 per cent of all maternal deaths due to anaemia (Ram Chandra, 2010).
2. Studies around the world have shown that lower birth weight is associated with an increased risk of a wide range of health problems in adult life including Type 2 diabetes, metabolic syndrome, chronic lung disease, osteoporosis and mental illness. There is also a wealth of data showing that poor

intrauterine and infant growth and nutrition are associated with reduced capacity in adult life, including reduced stature, lower physical work capacity, impaired cognitive function and educational attainment and (for women) an increased risk of low birth weight in the next generation (Ravelli *et al.*, 1998).

3. Energy and protein deficiency in mother is also associated with intrauterine growth retardation. Size at birth is strongly related to maternal mass index and during acute famine, birth weight falls by several hundred grams. Under nutrition of the mother, during her own foetal life and childhood growth, limits the growth of her foetus. Effect of the mother's current nutritional status is, therefore, influenced by her own past nutrition (Ram Chandra, 2010).
4. The Dutch famine studies have shown that maternal exposure to famine in late gestation was associated with glucose intolerance and insulin resistance in the offspring (WHO, 2001).
5. Past studies have also shown that, low maternal calcium intake is associated with lower bone mineral content in the children. Vitamin B-12 deficiency in the mother, especially if she was foliating replete, was associated with increased body fat and insulin resistance in the children (Gopalan, 1989).
6. Anaemia also makes a person more susceptible to carbon monoxide toxicity, which is one of the main pollutants in the biomass smoke (Ram Chandra, 2010).

Drought deteriorates the economic condition of every section of the society, residing in a drought affected region, and even to a certain extent of the people, living in a neighbouring area. The direct impact of it is on the food prices and production, which lately affects the weaker sections of the society and particularly the women. The same effect was felt in the Hathras district.

During the no drought situation, farmers in this region, involved in the production of wheat and pulses, employed women in the jobs of cleaning and sorting, in return of which they got *chanan* (lower part) and *phatkan* (upper part), this was utilized by these women for making *missi roti*, which served as a good source of protein and energy. Severe drought conditions reduced the work at farms to 50 per cent.

Under normal climatic conditions milk was available in plenty for all the sections of the society. People possessing milch animals offered buttermilk to poor people free of cost and even milk to pregnant women. The adverse effect of drought was that milk production came down and consumption of milk was far beyond the means of low income group. Buttermilk which was earlier available, free of cost was being sold during the drought at Rs. 5 per litre.

Pulses which are good source of protein for vegetarians have also become out of bounds. To add to the woes of common men, vegetables and fruits were also not available at reasonable rates during drought. For instance vegetables such as potato, tomato, spinach, cucumber, onion which were generally available at Rs. 6-10/kg, were being sold at Rs. 16-20/kg. Due to scarcity of water, leaves of radish become pale and dry and were of no use. Citrus fruits, serving as a good source of Vitamin C, proved to be out of reach of common people.

NFHS State Fact Sheet of Uttar Pradesh

Total percentage of anaemic population in India and in the Indian State of Uttar Pradesh, according to NFHS State fact sheet, is as follows :

	India	*Uttar Pradesh*
General	56.2	50.8
Urban	51.5	50.7
Rural	58.2	50.8
Uneducated	51.8	49.0
Educated	46.6	44.6

Though NFHS 3 data shows low iron status of general people in this region as mostly people are vegetarian, but still the health status of people is not bad, as they are able to get sufficient food through their earnings.

During my Intervention Programme for malnourished pregnant women, I found that the number of malnourished women suddenly increased during the drought. Hence, this study was designed to assess, whether this malnutrition is caused by the drought, and to investigate, up to what extent it has affected these womenfolk.

Materials and Methodology

150 pregnant women, registered with the District Women Hospital, were studied, during the months of July, August and September (2009). Government hospital was chosen for studying pregnant women, as in this hospital, pregnant women comes from nearby areas through ASHA (Accredited Social Health Activist) and majority of them belong to the economically weaker sections of the society. The targeted women were of second trimester, multipara, having one or two healthy babies, which were 50 in number. To obtain the desired information, a questionnaire was prepared after consulting doctors, subject experts and related books. The first section of the questionnaire was regarding general information, second was for clinical symptoms and visits to hospital and third was about dietary intake (48 hours recall method). Primary as well as secondary data from other sources were used for the study. Secondary data chiefly included, haemoglobin percentage collected from the Government hospital laboratory.

Results

Among the women surveyed in Government Hospital 80 per cent were homemakers, 12 per cent were industrial workers and 8 per cent were cottage workers (Fig. 6.1).

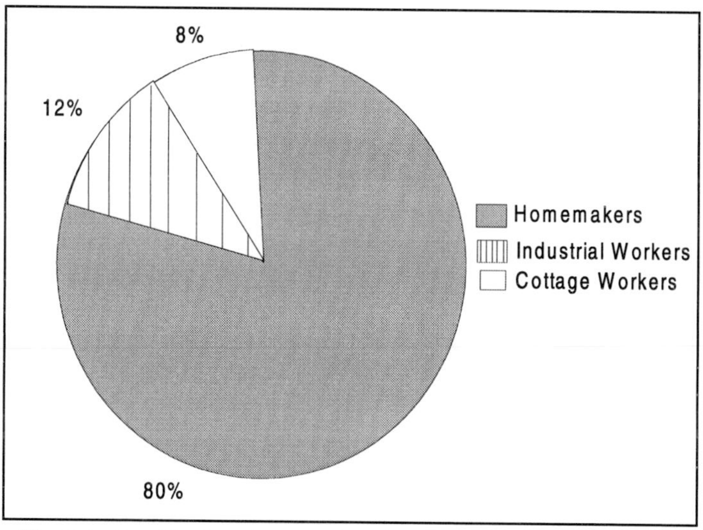

Fig. 6.1: Classification of women based on their occupational activities

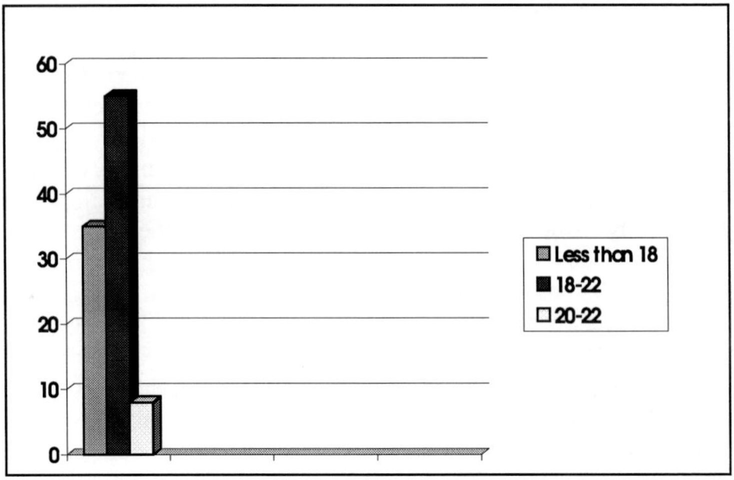

Fig. 6.2: Percentage distribution of BMI

According to ICMR, BMI (Body Mass Index) of pregnant women should be at least 22, but 36 per cent of the respondents had BMI less than 18, 56 per cent had less than 20 and only 8 per cent had BMI up to the desired level of 22 (Fig. 6.2.).

As per WHO classification, normal haemoglobin level for pregnant women of second trimester should be 11 gm/100 ml, Criteria for mild anaemia is 9.01-11 gm/100 ml, moderate anaemia is 7.01-9 gm/100 ml and for severe anaemia is 5.01-7 gm/ml. According to this criterion, among the women surveyed, 80 per cent were mild anaemic, 12 per cent moderate anaemic and 8 per cent were non anaemic (Table 6.1).

The daily consumption of foodstuffs by pregnant women, when compared to RDA (Recommended Daily Allowance) revealed that cereal consumption by 76 per cent of the respondents was in deficit.

Table 6.1: Percentage distribution of anaemia among the womenfolk studied

Anaemia	*Hb gm/100 ml*	*% of Women*
Normal	>11	8
Mild	9-10.9	80
Moderate	7-8.9	12
Severe	<7	—

Consumption of pulses was very poor, as 72 per cent of the respondents did not consume even 40 gm/day, against a recommended amount of 70 gm/day. Similar was the case with green leafy vegetables; 56 per cent of the respondents did not consume green leafy vegetables, even 25 gm daily. However, the intake of other vegetables, roots and tubers and fruits were fairly satisfactory, which was well above RDA. Majority of the pregnant women indicated banana as their fruit of choice. Deficit in milk consumption was found in 28 per cent women, which was nearly the RDA. Fat and sugar consumption were either up to the RDA or more than the desired level. Meat and egg consumption were also significantly less than the RDA (Table 6.2).

Table 6.2 : Consumption of daily foodstuffs by the pregnant women

	*RDA**		
Foodstuffs	*Veg (gm.)*	*Non-Veg (gm.)*	*% less than RDA*
Cereals	350	350	76
Pulses	7	45	72
GLV	150	150	56
Other vegetables	75	75	—
Roots and tubers	50	50	—
Fruits	30	30	—
Milk and milk products	325	225	28
Fat	30	30	8
Sugar	30	30	—
Meat, Fish and Egg	—	60	50

*Recommended Daily Allowances by ICMR.

Conclusion

The study showed that there was malnutrition among the pregnant women of second trimester, when there was drought only for a year. The situation would be more alarming when there will be a drought followed by heavy rainfall causing flood or a subsequent drought in winter, when wheat (main staple food of Uttar Pradesh) is sown. These abnormal climatic conditions will cause unavailability of sufficient food, resulting in more malnutrition, more LBW babies, more insulin resistant babies (diabetic) and more obesity, leading to creation of a vicious cycle. This creates dual burden on health and reduced life expectancy. Hence, we should innovate schemes and plans, and develop strategies in order to combat and prevent adverse climatic conditions, which could have disastrous impact on the general well-being of any section of the society.

REFERENCES

Gopalan, C.; Rama Sastry, B.V.; Balasubramaniam, S.C. 1989. *Nutritive Value of Indian Foods*. (Revised edition: Narsinga Rao B.S.; Deosthale Y.G.; Pant, K.C. editors). Hyderabad, India: National Institute of Nutrition (ICMR); 1989. http://bit.ly/5MOKb http://bit.ly/cqgNmL

Ram Chandra P. 2009. "Maternal and Child Nutrition: New dimensions of the dual nutritional burden." *Indian J. Med. Res*. 2009; 130:575-578.

Ravelli, A.C.J.; J.H.P. van der Meulen; R.P.J. Michels; C. Osmond; D.J.P. Barker; C.N. Hales; 1998.

WHO. 2001. *Iron Deficiency Anaemia Assessment Prevention and Control: A Guide for Programme Managers*. Geneva, World Health Organization.

7
The Psychology of Global Warming and Climate Change

Ali Seid Mohammed and *Uma Melakania*

Introduction

As we move into the twentyfirst century, we do so in a world of significant technological progress, industrialization, and globalization. Our current standard of living is based upon these drivers of modern progress. The agricultural sector, through mechanization and other technological and biological advances, has been able to supply our growing sustenance requirements. Telecommunication systems have enabled friends, families, businesses, organizations, and governments to communicate verbally, as well as visually, across thousands of miles. Combine this with our modern transportation systems, and we remove the notion of the frontier. Even the deepest oceans and other planets are now being explored by scientists. Having achieved such enormous progress, why then should we be concerned about the future of humankind on a global scale?

The answer to this question is quite simple; if we just think only over one billion people have any meaningful access to the resources and quality of life described above. The rest of the world's population—some five billion people who live in less developed regions—have only a taste of what this lifestyle might be like (UNDP, 2003). Such inequality is leading towards a world of growing disparity between industrialized and developing nations. In effect, the communities of less developed regions are held captive to the needs and wants of industrialized nations, whose living and consuming habits are in many ways condemning billions of people to a lower quality of life.

It is clear that a global drive to reach the westernized view of the good life, without a drastic change in production processes and consumption patterns, will soon bring us up against ecological limits, forcing us to rethink what we mean by a secure and fulfilling lifestyle. However, as Wilson (2002) puts it, if each people of the globe were to live the lifestyle of the average American, holding technology constant, we would need resources of four more Earths.

These presented a highly simplified and polarized view of the world in which there are those who *have* and those who *have not*. While this might well be the case at the extremes, there is clearly a continuum of positions in between. The central argument of this article is that if we are interested in the well-being of current and future generations, we should not only be concerned for the future of the world, we should be actively searching for new ways to enable individuals, communities, and nations to live a sustainable life. Both the extreme lifestyles are not good for healthy vital functioning of human being.

By understanding the psychology of human behaviour, it will be possible to develop strategies/programmes that will not only be more effective in communicating the vision of sustainable development to the public, but will assist society in adopting new behaviours that will help realize a more sustainable future (McKenzie-Mohr, 2002). It is to say that devising sustainable strategies without taking the psychological basis of behaviour into account is not likely to be successful. Let us first see the history of global climate concerns as a background for subsequent psychological discussions.

The Emergence of a Concern for Global Climate Change

The 1970s witnessed the early formation of the fourth environmental driver of sustainable development—the concern that greenhouse gases from anthropocentric sources are leading to a disruption of the global climate. However, during the 1970s, the catalyst for this concern did not begin with a focus on greenhouse gases *per se*, but with the human induced destruction of ozone in the stratosphere and the resulting increased risk of cancer from the increased intensity of ultraviolet rays.

The early debates about the ozone layer in the 1970s exposed a much broader question of whether human activities in general were affecting the global climate. The first World Climate Conference marked the beginning of worldwide concern for global climate change, the fourth environmental driver of sustainable development. In 1979, the World Meteorological Organization (WMO), after issuing an authoritative statement on global climate change in 1976, convened the First World Climate Conference in Geneva, Switzerland. The purpose of the meeting was to share scientific knowledge and data on the causes of climate change and to identify the potential impacts that climate change might have upon human activities. The conclusions of the conference were presented in the Declaration of the World Climate Conference, which "*highlighted the international community's emerging perception of the climate as a vital natural resource*". The Declaration stated that it is urgently necessary for nationstates around the world "*to foresee and to prevent potential manmade changes in climate that might be adverse to the well-being of humanity*". (Figueres, 2002)

The Declaration furthermore, identified increasing atmospheric concentrations of CO_2 from the combustion of fossil fuels, deforestation, and changes in land-use as the leading cause of global climate change. The Conference led to the formation of the World Climate Programme (WCP) that was to be jointly administered by the WMO, UNEP, and the International Council of Scientific Unions (ICSU) to improve understanding of the climate system.

Interestingly, while ozone depletion and global climate change were initially seen as separate issues, the international protocols developed (or amended) for both issues during the 1990s began to formally link the two problems. For example, PFCs (perfluorocarbons) and HFCs (hydrofluorocarbons) were treated as greenhouse gases by the 1997 Kyoto Protocol and as Ozone Depleting Substances (ODS) in the Montreal Protocol (Malabed *et al.*, 2002). Similarly, while CFCs and HCFCs (hydrochlorofluorocarbons) have global warming potential, they are not covered by the Kyoto Protocol since they are adequately addressed by the Montreal Protocol (UNEP, 1999). In addition, both Protocols recognize the importance of HCFCs as "*transitional*" substances that play a crucial role in phasing out ODS.

However, the potential impact of the Kyoto Protocol on the international economy was far greater than that of the Montreal

Protocol. Similarly, whereas the Montreal Protocol controls consumers and producers, the Kyoto Protocol focuses on the controls of emissions (or containment), which is much more difficult to manage due to the scale of greenhouse gas emissions. These differences make integrating the objectives of the protocols more challenging, leading to a call for more "*flexible approaches*" to potential solutions to global climate problems (UNEP, 1999). Regardless of the difficulties faced when trying to implement and integrate the Montreal and Kyoto Protocols, both Protocols signal that concern for the global climate has become a major component of the international environmental—and, hence, sustainable development—agenda.

The historic U.N. climate talks in Copenhagen (2010) ended after a 31-hour negotiating marathon, narrowly avoiding the collapse by accepting a compromise to give billions to poor nations to deal with global warming but does not require the world's major polluters to make deeper cuts in their greenhouse gas emissions. Two weeks of wrangling exposed sharp divisions between rich and poor nations (the haves and have nots)—and even among major greenhouse CO_2 emitters like China and the United States—did not agree on how to fight global warming. The representatives in the conference span reflect the pressure behind from their national interest and people's different perception of global warming/climate change.

The Psychology of Human Needs

What you feel, understand, think, and do now is based upon your subjective view of realities, which is shaped by factors such as experience, education, and societal and cultural values. An interesting concept to consider in the following discussion is that your personal consciousness will help you decide whether you agree or disagree, even with the ideas being put forward. Hence, your interpretation of this chapter is likely to be different from someone else's. The desire to understand human consciousness and its relationship to our behaviour is the motivation for the field of psychology, which is the science of behaviour and mental processes.

While the complexity of this topic makes it difficult to formulate universally accepted ideas, during the past century psychologists have been able to develop theories that can help us understand why people might be behaving the way they do. There is also a strong motivation to look at humans through needs, since it "enables us to build a bridge between a philosophical anthropology and a political option". (Max-Neef *et al.*, 1989)

The concept of human needs has traditionally been used in two distinct ways (Ryan, 1995). The first and most common definition of a need is connected to virtually any *motivating force*, such as one's desires, goals, wants, or values. When used in such a way, an individual is using the term 'need' to express a strong desire. The second definition of need refers to *physiological needs*; the nutrients or conditions which are essential for our survival as humans, such as food, water, and shelter. Using these two definitions, it is possible to generate endless lists of needs, especially if the loose definition of a need is invoked.

However, if the criterion of a need is focused on necessity for growth, the potential list of needs is rapidly reduced. Human needs or desires motivate behaviour. The concept of *motivation* grouped into four broad areas (Boeree, 2006) :

1. *Biological Motivations*—Organically driven, such as our need for food and water, and hedonistic needs such as pain-avoidance and pleasure;
2. *Social motivations*—The need for acceptance, attention, and approval which influence forms of self-esteem;
3. *Personal Motivations*—Based upon the experiences of the individual resulting in habitual behaviour, defense mechanisms, and personality styles; and
4. *Higher Motivations*—Which come in two forms :
   - *Competence motivation*—A desire to learn, attain competence/mastery, and be creative; and
   - *Altruistic motivation*—A concern for society, and a need for compassion and love.

The failure to differentiate the various types of human needs has created confusion in the realm of politics, especially when it comes to discussions about sustainable development. While it is possible to scientifically define biological motivations (our need for food and water, etc.) the other types of motivation cannot easily be defined. The problem is further compounded by the fact that within the field of psychology, behaviourists, psychoanalysts, and humanists. They approached the topic of human needs from different perspectives. Hence, human needs are discerned differently according to the ideological and disciplinary lens of the view (Max-Neef *et al.*, 1989).

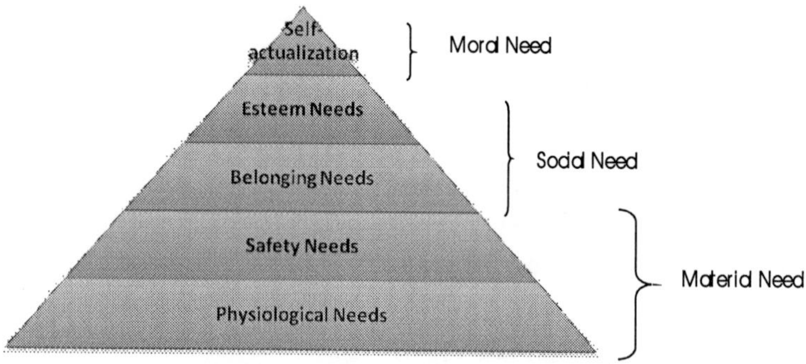

Fig. 7.1 : Maslow's Hierarchy of Needs

Maslow (1943) developed a theory of human motivation which can be used to analyze the *functioning individual.* Maslow's theory consists of a hierarchy of needs (Figure 7.1) and is based upon the notion that as an individual's lower level needs satisfied, new needs emerge that require fulfilment.

As with a useful way of thinking about how environmental and societal factors might influence an individual's behaviour, it is especially important when considering differences in the needs of people in developed *versus* developing nations. In addition to using Maslow's hierarchy of needs to assess humans, researchers have applied this theory to a variety of research areas. An interesting example is a paper by Collins *et al.* (1997) which expands Maslow's hierarchy of needs into a framework for approaching the human dimension in ecosystem management.

In contrast to Maslow's theory, Max-Neef *et al.* (1989) argue that all human needs must be understood as an interrelated and interactive system. Apart from sustenance needs, which are essential to human life, there are *no* hierarchies in this system. They question the lack of any real discussion of the difference between human *needs* and the *satisfiers* of those needs. In particular, they ask *what* the fundamental needs are and *who* determines them? To help answer this question, they present a 36 cell matrix which organizes human needs into "existential (*being*, *having*, *doing*, and *interacting*) and axiological (*subsistence*, *protection*, *affection*, *understanding*, *participation*, *idleness*, *creation*, *identity*, and *freedom*) needs (Table 7.1)." Within this framework, needs are seen as having a two-fold character: as deprivation and as potential.

From this framework, Max-Neef *et al.*, (1989) make the following two claims. "*First: Fundamental human needs are finite, few, and classifiable. Second: Fundamental human needs ... are the same in all cultures and in all historical periods. What changes, both over time and through cultures, is the way or the means by which the needs are satisfied*".

What Max-Neef *et al.*, (1989) argues is that it is the economic, social, and political *systems* which adopt different methods for the satisfaction of the same human needs. Hence, a critical aspect to their theory is that individuals, groups, and cultures develop the *actions* and *value systems* that will be used to realize their needs. Understanding and describing these value systems is ultimately subjective, which presents a problem since the Western view of development tends to be based upon objectivity, i.e., the view that we should deal with facts or conditions without distortion by personal feelings or interpretations.

They also argue that social and economic relations, which are defined by historical and cultural circumstances, need to be viewed with both subjectivity and objectivity. Hence, "*satisfiers are what render needs historical and cultural, and economic goods are their material manifestation*". An important outcome of the above discussion is that individuals, communities, and societies decide upon the satisfiers to their needs. This is critical since it enables us to reinterpret the concept of poverty, which has traditionally been defined in economics with reference to income. Thus, they said "*we should speak not of poverty, but of poverties; in fact, any fundamental human need that is not adequately satisfied reveals a human poverty*" (Max-Neef *et al.*, 1989).

It follows that poverty can be found in all nine of the axiological needs, e.g., poverty of sustenance, of protection, etc. These poverties lead to confusion, isolation, frustration, and general psychological distress, which ultimately manifest themselves through psychopathic such as war, racism, hunger, etc. The insights into human condition contain a valuable message. If we rely only on an economic frame of reference, we will ultimately not be able to satisfy our needs. Therefore, our research, strategies, and policies need to be developed using a *trans-disciplinary* approach.

What are the Psychosocial Impacts of Climate Change?

The psychosocial and mental health implications of climate change

Table 7.1: A matrix of existential and axiological human needs

Fundamental Human Needs	*Being (Qualities)*	*Having (Things)*	*Doing (Actions)*	*Interacting (Settings)*
Subsistence	Physical and mental health	food, shelter, work	feed, clothe, rest, work	living environment, social setting
Protection	Care, adaptability, autonomy	social security, health systems, work	co-operate, plan, take care of help	social environment, dwelling
Affection	Respect, sense of humour, generosity, sensuality	friendships, family, relationships with nature	share, take care of, make love, express emotions	privacy, intimate spaces of togetherness
Understanding	Critical capacity, curiosity, intuition	literature, teachers, policies, educational	analyse, study, meditate, investigate	schools, families, universities, communities,
Participation	Receptiveness. dedication, Sense of humour	responsibilities, duties, work, rights	cooperate, dissent, express opinions	associations, parties, churches, neighbourhoods
Leisure	Imagination, tranquillity, spontaneity	games, parties, peace of mind	day-dream, remember, relax, have fun	landscapes, intimate spaces, places to be alone
Creation	Imagination, boldness, inventiveness, curiosity	abilities, skills, work, techniques	invent, build, design, work, compose, interpret	spaces for expression, workshops, audiences
Identity	Sense of belonging, self-esteem, consistency	language, religions, work, customs, values, norms	get to know oneself, grow, commit oneself	places one belongs to everyday settings
Freedom	Autonomy, passion, self-esteem, open-mindedness	equal rights	dissent, choose, run risks, develop awareness	anywhere

Source : *Max* - Neef *et al.* (1989).

have gained attention in the context of disaster recovery from extreme weather events (Few, 2007). Fritze and colleagues (2008) note that direct impacts, such as extreme weather events, are likely to have immediate effects on the prevalence and severity of mental health issues in affected communities as well as significant implications for mental health services; vulnerable communities will experience ongoing disruptions to the social, economic and environmental determinants that promote mental health in general; and finally climate change as a global environmental threat may create emotional distress and anxiety about the future.

Emotional reactions are critical components of information processing and also have a direct relation to physical and psychological health (Dillard and Pfau, 2002 in Moser, 2007; Groopman, 2004). It is hypothesized that certain strong emotional responses such as fear, despair, or a sense of being overwhelmed or powerless can inhibit thought and action (Macy & Brown, 1998; Moser, 2007; Nicholson, 2002). As Moser and Dilling (2004) illustrated, well-meaning attempts to create urgency about climate change by appealing to fear of disasters or health risks frequently lead to the exact opposite of the desired response: denial, paralysis, apathy, or actions that can create greater risks than the one being mitigated. Moser (2007) has reviewed the emotional responses to informational messages about climate change.

The Public Perception of Global Warming

Several attempts have been made in various countries assessing public views of global warming and/or climate change problems. Several of the papers using survey methodology are concerned with this basic question and address the determinants of public opinion on global warming and public support for environmental policies (Krosnick *et al.*, 2000). Another issue is how well-established, heuristics, and biases in decision-making affect thinking about global warming (Gilovich *et al.*, 2002).

According to Oppenheimer and Todorov (2006) one of the problems that US policy-makers face in attempting to implement effective environmental policy is that environmental issues have become politically laden. In 1997, in an effort to increase public support for the Kyoto agreement, the Clinton administration sponsored an extensive media campaign about the scientific evidence and consequences of global warming. National surveys were also conducted before and after the campaign and it showed little change overall in public opinion (Krosnick *et al.*, 2000). In fact, both Republicans and Democrats have changed their opinions but in opposite directions. Republicans become less concerned about global warming whereas Democrats become more concerned. Citizens used the source of the informational campaign to draw inferences about the value of the information.

As noted earlier, there seems increasingly higher level of public concern about global warming in Europe and the US. However, the personal relevance of the concern is not strong. As Leiserowitz (2008) shows, only 13 per cent of Americans were concerned about the impact of global warming on themselves and their family or on their local community. These findings coincide with Weber's argument that people are likely to act on decisions derived from affective feelings and personal experience but from statistical descriptions of risks. Taken together, these studies point to additional difficulties on the policy front until the impacts of global warming is broadly visible. Also relevant are the high public support for governmental interventions that do not directly affect citizens' individual behaviours and their actions (Leiserowitz, 2008). Thus, the public in both Europe and the USA might have been 'concerned but not moved'.

Human perceptions and judgments about climate change are important because they affect levels of concern and, in turn, the motivation to act. Public opinion data in the United States indicate that climate scientists are more concerned about the possibly severe effects of climate change on human populations, ecosystems, and infrastructures than average citizens and governmental officials (Dunlap & Saad, 2001). One national representative opinion poll shows that 47 per cent of people polled view global warming as a "very serious" problem, and another 28 per cent view it as a "somewhat serious" problem (Pew Project, 2006). This level of concern—75 per cent of people in the U.S. assessing global warming as a "very" or "somewhat" serious problem—is similar to the level in Russia (73%) and lower than that in many other nations: 87 per cent of Canadians, 81 per cent of Mexicans, 95 per cent of French, 88 per cent of Chinese, 97 per cent of Japanese, 96 per cent of Brazilians, and 94 per cent of Indians assess global warming as a "very" or "somewhat" serious problem.

The lack of knowledge of political and scientific facts can

combine with decision biases to further undermine successful policy actions. Wariness about global warming can reflect narrow concerns about local temperatures rather than reasoning about its large scale consequences. As pointed out by Lorenzoni and Pidgeon (2005), the degree of concern in European countries seems to vary as a function of the local temperature. In fact, the degree of concern reported for 15 European countries, is strongly correlated with the average temperature in July, the hottest month in Europe. On the other hand, temperature is only one factor in the mind of both people and governments. Other impacts, such as incidence of strong storms, also affect both policy-makers and the general public. And, with Germany and Scandinavian countries having historically taken a leading role on this issue, the relation between temperature and national policy on global warming is at best obscure.

Denial of Climate Change

Scientific evidence shows the main influences of climate change are behavioural—population growth and energy consumption. "What is

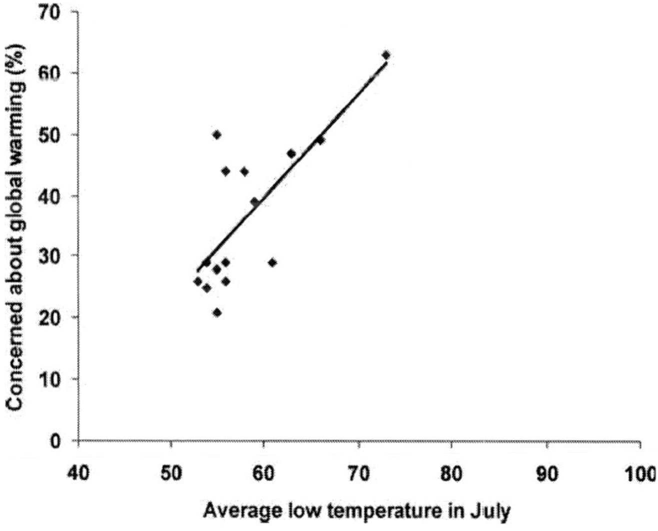

Fig. 7.2 : Scatter plot of the average daily minimum temperature (°F) in July and the percentage of respondents reporting that they are "very worried" about climate change for 15 European countries (Lorenzoni and Pidgeon, 2005)

unique about current global climate change is the role of human behaviour, we must look at the reasons people are not acting in order to understand how to get people to act" (APA, 2010).

According to Keim (2009) even as the science of global warming gets stronger, fewer Americans believe it is real. In some ways, it is nearly as jarring disconnect as enduring disbelief in evolution or carbon dating. According to Kari Marie Norgaard, a Whitman College sociologist who has studied public attitudes towards climate science, we are in denial. "*Our response to disturbing information is very complex. We negotiate it. We do not just take it in and respond in a rational way*", said Norgaard (2010).

So we pretend it is not a problem, but we do not want to make it seem insensitive. Sometimes people who are very empathetic are less likely to help in certain situations, because they are so disturbed by it. The human capacity of empathy is really profound, and that is part of our weakness. If we were more callous, then we would approach it in a more straightforward way. It may be a weakness of our capacity as sentient beings to cope with this problem.

Denial is not the only explanation and psychological barrier, Dany Gilbert (2007), an American psychologist, on youtube.com tried to explain why the predictive human brain failed to help us. He said, ". . . *When you hold your fate on your hands, why would you ever make a fist*?" and tried to explain it in an evolutionary context. As human brain is highly evolved to detect and react instantly to threats which have accustomed features such as human face (signs) and intentional actions (Anthrax than Influenza); violating our moral values (e.g., Food and Sex); and others for which human adapted for millennia and our brain has failed to respond because global warming lacks those key features that trigger us into action. Thus, there seems evolutionary explanation as a reason why people not respond to global warming. Whether because of our brain evolution or defense reaction, denial is not the only barrier but there are other barriers to be discussed in the next section.

Attitude as a Root Cause

The drastic increase in the emission of CO_2 (carbon dioxide) within the last 30 years caused by burning fossil fuels to fulfil the ever expanding human need described and classified by different psychologists. Our need for burning fossil fuels is identified as the major reason for the change of temperature in the atmosphere and result in global climate change. However, it took more than 20 years to broadly accept that mankind is causing global warming with the emission of greenhouse gases. In addition to the ever expanding needs, our attitude is major psychological factor that prevents us to think and act globally.

An attitude is a hypothetical construct that represents an individual's degree of like or dislike for an item. Attitudes are generally positive or negative views of a person, place, thing, or event which is often referred to as the attitude object. People can also be conflicted or ambivalent toward an object, meaning that they simultaneously possess both positive and negative attitudes toward the item in question. Attitudes are judgments that develop on the ABC model (affect, behaviour, and cognition). The affective response is an emotional response that expresses an individual's degree of preference for an entity. The behavioural intention is a verbal indication or typical behavioural tendency of an individual. The cognitive response is a cognitive evaluation of the entity that constitutes an individual's beliefs about the object. Most attitudes are the result of either direct experience or observational learning from the environment (Wikipedia, the free encyclopedia online). It is our attitude (whether right or wrong) determines our behaviour and actions.

We know that our energy demand is simply too high. More than 80 per cent of the world-wide energy demand is currently supplied by the fossil fuels: coal, oil or gas. It took nature millions of years to generate these resources, however we may have used them up within the next decades. Even, the shrinking supply will not make it possible to continue as usual for a longer time, the dominant public attitude is still 'it will be impossible to find alternative sources in a short or medium term.' What about the balance of nature that regulate global climate. The general attitude still is we cannot afford changing our behaviour but we can ignore or justify our action. Thus, the main cause of global warming is our treatment of "Nature" and the root cause is our thoughtless attitude to Nature. This may be a simple explanation for the reason :

- Why the warnings about climate change been ignored for more than 20 years?
- Why were ever more scientific evidence demanded to find

the coherence of man-made CO_2 emissions as cause of global warming? Why was not common sense reason enough to act?

- Why do most people refuse to change their personal behaviour voluntary in order to reduce CO_2 emissions caused by their activities?

It would be of great interest to revisit some studies now. Recently American Psychologist Association (APA) Task Force discussed the Interface Between Psychology and Global Climate Change; examined decades of psychological research and practice that have been specifically applied and tested in the arena of climate change, such as environmental and conservation psychology and research on natural and technological disasters. It cites a national Pew Research Centre poll in which 75 per cent to 80 per cent of respondents said that climate change is an important issue. But respondents ranked it last in a list of 20 compelling issues, such as the economy or terrorism. Also, some studies have looked at whether financial incentives can spur people to weatherize their houses.

After the devastating European heat wave of 2003, the Atlantic hurricane seasons of 2004 and 2005, the media coverage of the consequences of Arctic warming was great. Then, the public now view global warming in more personal terms. How pervasive is this view, has it affected attitudes toward particular policies, and how long may it last? According to the American Psychological Association (APA, 2010) conclusion that research in cognitive psychology suggests that certain perceived characteristics of climate change are "natural", not new, and in principle they are controllable. It may lead citizens as well as policy-makers to underestimate the magnitude of the risks. Other psychological research provides additional hypotheses related to emotional reactions to climate change such that the absence of feeling at risk may be a psychodynamic reaction, the result of psychic numbing or denial in the face of overwhelming and uncontrollable risk.

Other Psychological Barriers to Actions

Although some people in every society are changing their behaviour in response to climate change, humans in the aggregate continue to produce greenhouse gases in quantities that drive further change. There are number of important psychological barriers or obstacles that may hinder behaviour change that would be adaptive or would help to mitigate climate change. Many of the psychological barriers to climate change behaviours are less well documented than the structural ones. However, evidence for many of them exists in the context of other pro-environmental behaviours, so they may also operate in relation to limiting climate change.

According to APA (2010) the psychological obstacles to adequate climate-change adaptation can be listed and sequenced by psychologists. It begins with genuine ignorance through increasing awareness and intention to act which, however, are hampered by one psychological process or another, to tentative, incomplete adaptation choices that are subject to social and other obstacles. The complete list is :

1. *Ignorance*: Unaware or in denial.
2. *Uncertainty*: Unsure of the facts or what to do, do not trust experts or believe their conclusions.
3. *Mistrust and reactance*: Distrust risk messages.
4. *Judgmental discounting*: Undervaluing future or distant risks.
5. *Place attachment*: More likely to care for a place to which they feel attachment than for one they do not.
6. *Habit*: Behavioural momentum because they are fixed in their ways.
7. *Perceived behavioural control*: Think that it is a global problem and individuals can do nothing about it.
8. *Perceived risks from behavioural change*: Believe that others should act, or believe that their actions will make no difference or are unimportant compared to those of others.
9. *Tokenism and the rebound effect*: Engaging in token actions or actions they believe are helpful but objectively are not.
10. *Social comparison, norms, conformity, and perceived equity*: Routinely compare their actions with those of others, and derive subjective and descriptive norms from their observations about what is the "proper" course of action.
11. *Conflicting goals and aspirations*: Other worthy goals and aspirations draw their time, effort, and resources.
12. *Belief in solutions outside of human control*: Believe that external factors beyond human actions or control will address the problem.

Besides identifying a dozens of barriers presented previously, the task force highlighted some ways that psychology is already working to limit these barriers. For example, people are more likely to use energy-efficient appliances if they are provided with immediate energy-use feedback. Devices that show people how much energy and money they are conserving can yield energy savings of 5 per cent to 12 per cent. According to research, "Behavioural feedback links the cost of energy use more closely to behaviour by showing the costs immediately or daily rather than in an electric bill that comes a month later".

Summary

Social participation is a fundamental part of the sustainable development process as emphasized at the Rio 92 and Johannesburg 2002 Conferences. Sustainable development is a social paradigm, and it is from society that changes towards greater sustainability will be accomplished. The sustainable development process involves people and groups of people that have basic needs for a better quality of life. Human being has an expanding, complex and culturally attuned needs and need satisfiers. This has created the so-called modern technology world of affluence.

In our technology and scientific minded world, we seem to have forgotten that mankind is only a relatively minor part of Nature. Our perception is to be able to control Nature instead of trying to arrange ourselves within Nature. Our attitude not only ignore that we are being part of a larger world, but also try to deny the great risk of changing global climate. In addition, our unit of measure is more and more often only money and what has no price tag, seems to have no value to us any more. In doing so, we mix up economic growth with general well-being and financial income with personal happiness.

Climate change, adaptation, and mitigation responses have the potential to have different impacts on different populations. Further, different populations may have different social constructions of the meaning of climate change and its mitigations. Different cultural meanings and social justice concerns are important to address in order to recognize the limits of one's research and to be able to speak to and be heard by different populations. Further, attending to social and cultural differences can potentially further suggests new ways of thinking about basic psychological processes and new ways of addressing climate change.

However, sustainable development is a collective response based on individual commitments to contribute to this process at the local level. The participation of stakeholders' in social participation schemes provides the opportunity for people to express their opinions and feel part of the planning process. It is important to stress that everyone has an important role to play in helping to protect the environment and improve measures to combat climate change. Thus, psychologists are urged to continue to expand the knowledge base so that barriers for action should be removed on time.

REFERENCES

American Psychological Association. 2010. Psychology and Global Climate Change: Addressing a Multi-faceted Phenomenon and Set of Challenges. A Report by the American Psychological Association's Task Force on the Interface between Psychology and Global Climate Change.

Boeree, G. 2006. The Ultimate Theory of Personality, Motivation. www.ship.edu.

Collins, T.K.; J.R. Craig and D.B.K. English. 1997. Meeting Human Needs for Food, Water, Minerals and Other Basic Resources: Maslow's Hierarchy of Human Needs and its Foundation. *General Technical Report*, SRS 17, Southern Research Station, USDA-Forest Service.

Dany Gilbert. 2007. Why human brain fail to react for the threat of Global Warming. Dany gilbert@POP!tech 2007. Video clip online. http://www.youtube.com

Few, R. 2007. Health and climatic hazards: Framing social research on vulnerability, response and adaptation. *Global Environmental Change*, 17, 281-295.

Figueres, C.E. 2002. *Establishing National Authorities for the CDM: A Guide for Developing Countries*. International Institute for Sustainable Development and the Centre for Sustainable Development in the Americas.

Fritze, J.G.; G.A. Blashki; S. Burke, and J. Wiseman. 2008. Hope, despair and transformation: Climate change and the promotion of mental health and well-being. *International Journal of Mental Health Systems*, 2(13). Retrieved April 30, 2009, from http://ijmhs.com/content/2/1/13.

Gilovich, T.; D. Griffin and D. Kahneman (eds.). 2002. *Heuristics and Biases: The Psychology of Intuitive Judgment*. Cambridge University Press.

Groopman, J. 2004. *The Anatomy of Hope*. New York: Random House.

Keim, B. 2009. The Psychology of Climate Change Denial. http://www.wired.com

Krosnick, J.A.; A.L. Holbrook and P.S. Visser. 2000. 'The impact of the fall 1997 debate about global warming on American public opinion, *Public Understanding of Science*, 9, 239-260.

Leiserowitz, A. and K. Broad 2008. Florida: Public opinion on climate change. A Yale University/University of Miami/Columbia University Poll. New Haven, CT: Yale Project on Climate Change.

Lorenzoni, I.; N.F. Pidgeon and R.E. O'Connor, 2005, 'Dangerous climate change: The role for risk research', *Risk Analysis* 25, 1387-1398. DOI. 10.1111/j. 1539-6924.2005.00686.x

Macy, J. and M.Y. Brown. 1998. *Coming Back to Life: Practices to Reconnect Our Lives, Our World*. Gabriola Island, British Columbia: New Society Publishers.

Malabed, L.; J. Velasquez, and R.E. Shende. 2002. *Inter-linkages between the Ozone and Climate Change Conventions: Part I—Inter-linkages between the Montreal and Kyoto Protocols*. United Nations University, United Nations Environment Programme/Division of Technology, Industry and Economics (DTIE), Massachusetts Institute of Technology Global Accords Programme and the Alliance for Global Sustainability/ Value of Knowledge Project.

Maslow, A.H. 1943. "A Theory of Human Motivation." *Psychological Review*, 50, 370-396.

Max-Neef, M.; A. Elizalde and M. Hopehayn. 1989. "Human Scale Development: An Option for the Future. *Development Dialogue*, 1, 5-81.

McKenzie-Mohr, D. 2002. "The Next Revolution: Sustainability." *Psychology of Sustainable Development*, P.A.S. Schmuck, W.P. (ed.). Kluwer Academic Publishers, Boston, 19-36.

Moser, S.C. 2007. More bad news: *The Risk of Neglecting Emotional Responses to Climate Change Information*. In S.C. Moser and L. Dilling (eds.), *Creating a Climate for Change*. New York: Cambridge University Press.

Moser, S.C., and L. Dilling 2004. "Making climate hot: Communicating the urgency and challenge of global climate change." *Environment*, 46, 32-46.

Nicholsen, S.W. 2002. *The Love of Nature and the End of the World*. Cambridge: MIT Press.

Norgaard. 2010. http://www.whitman.edu/content/sociology/faculty/norgaard

Oppenheimer, M. and A. Todorov. 2006. Global warming: The Psychology of Long Term Risk (Guest Editorial). *Climatic Change*, DOI: 10.1007/ s10584-006-9086-6 c_ Springer.

Ryan, R.M. 1995. "Psychological Needs and the Facilitation of Integrative Processes." *Journal of Personality*, 63:3(September).

United Nations Development Programme (UNDP). 2003. Human Development Report. *Millennium Development Goals: A Compact among Nations to End Human Poverty*. Oxford University Press, New York.

United Nations Environment Programme (UNEP). 1999. *Round Table Discussion, Climate Change and Ozone Protection Policy*: Two Protocols—One Response. United Nations Environment Programme Division of Technology, Industry and Economics (TIE) and Ozone Action Programme.

Wilson, E.O. 2002. *The Future of Life*. Alfred A. Knopf, New York.

PART II

FROM THE CLIMATE OF DESPAIR TO THE CLIMATE OF HOPE

Existing Potentials to Respond Maldevelopment

8

Forest Based Human Security Source for the Indian Tribal Population

Rita Singh Raghuvanshi and *Nandana Kumari*

Introduction

Cooling the Earth, at present time, is the greatest concern and challenge before the humanity. There is urgent need of climate change mitigation efforts for the very survival and sustainability of our planet. Now, it is right time for us to evolve all possible strategies which could be of critical value for resolving these vital issues. This demands efficient environmental management with focus on carbon emission reduction, enhanced carbon sequestration, increased use of clean and non-polluting sources of energy that could be urgent for evolving eco-centric strategies that could lead to cooling of the Earth. By analysing these points, the ultimate solution to mankind in this aspect is forest.

Global Impacts of Forest Degradation

Forest is very essential to maintain our environment safe and sound. As deforestation and forest degradation leads to higher rates of soil erosion and siltation of waterways, loss of species and genetic diversity and an increase in carbon emissions, all these contribute to global warming (Lipper, 2005).

Forest as an Environment Security Source

Forest is world's air-conditioners and earth's blankets. Without forests, this world would be an inhospitable place to live in. Desired level of ecosystem maintenance can vary significantly depending on the forest management. Two important services provided by forest ecosystem which benefit the global community are carbon sequestration and storage and the conservation of biological diversity through the provision of habitat for highly diverse plant and animal species (Prasad, 2007). Global climate change is associated with rising levels of greenhouse gases (in particular carbon dioxide) in the atmosphere. Forest ecosystem, including above and below ground complements, are major carbon sinks, taking up carbon from the atmosphere, thus they have an important role in mitigating climate change (Lipper, 2005).

Forests and India

India is the second most populous and seventh largest country in the world with a forest area approximately 23.68 per cent of its geographical area. Forestry is the second largest land-use after agriculture. Forest remain from the beginning, very good source of foods, fodders, fuel, timber and non-wood forest produce for the forest dwellers as well as for the rural communities living in and nearby forest areas. Out of 0.6 million villages in India, around 0.2 million villages with a population of about 350 million are dependent for subsistence directly or indirectly on the forest (Malik *et al.*, 2010).

Jharkhand : An Introduction

Jharkhand means "Land of Forest". It is a young State in Eastern India, established on 15th November, 2004. It shares its borders with five States, Bihar in North, Uttar Pradesh and Chhattisgarh in West, Orissa in South and West Bengal in East. Its approximate 74 per cent of the total population resides in rural areas. Approximate 54 per cent population lives below poverty line. According to 2001 Census, 26.3 per cent of the total population is ST population. The Scheduled Tribes are primarily rural and about 91.7 per cent of them reside in villages. The overall sex ratio of the ST population in Jharkhand is 987 females per 1000 males which is higher than the national average (978) for the total ST population. The most important thing, Jharkhand is leading producer of minerals or the richest State in the natural resources wealth in the country (www.jharkhand.nic.in.).

The Tribes of Jharkhand

India has a variety of tribal population, reflecting its great ethnic diversity. Tribes constitute an important segment (approximately 8%) of the Indian population. In Eastern India, majority of the tribes reside. Most of the tribal population in India lives in forest ecosystems and has its own kind of socio-cultural pattern, traditions and typical food practices (Sinha and Lakra, 2007).

The Jharkhand State has 30 types of tribes and 8 types of Primitive tribes. Main tribes are Oraon, Munda, Santhal, Ho and Gond, etc. The other tribes are Bedia, Karmal, Lohra, Bhumij, Kharia, Chero and Kharwar (www.jharkhand.nic.in). Primitive tribes are Asur, Parhaiya, Savar, Birhor, Birjia, Korwa, Malpaharia and Sauria Paharia.

Characteristics of the Major Tribes

The tribes of Jharkhand are ranging from hunter-gatherers, forest dwellers to settled agriculturist tribes like Santhal, Ho, Munda, Oraon and Gond. Even today, the agriculturist tribes still engage in

Fig. 8.1: Jharkhand State

a considerable amount of gathering activity and there remains evidence of collectivist behaviour in economic activities (Malik, Kaushal and Sah, 2010). Most of these tribals have small or marginals landholdings. They grow foodgrains for 8-10 months, however, for rest of the month, they depend on the forest for their all kinds of need, whether food, medicine or livelihood need (Sinha and Lakra, 2007).

Characteristics of the Primitive Tribes

They solely depend on the uncultivated and forest plants for their requirement from food to medicine need. They hardly come out of the forest to the plain area, only to collect salt, potato, or onion in exchange of honey and resins collected by them. They have been conserving plant and crop genetics resources as well as the knowledge on their utility without expecting any return (Rout, 2010).

Agriculture in Jharkhand

Agriculture in the Jharkhand is characterised by low productivity of foodgrains and cash crops, due to its meagre assured irrigation facility. Jharkhand's mono-cropped type of farming system and its undulating topographical features also lead to poor production. A large number of landholdings are uneconomic. Another reason for underdeveloped character of agriculture is predominance of food crops (92% of the total cropped area) over cash crops (3-5% of the total cropped area).

Agriculture is the main source of livelihood for the Jharkhand people. About 70 per cent of the people depend on small farmlands. To meet the ever growing need of increasing population, there is least scope of horizontal expansion of the land for agricultural purpose. In this type of situation, forest will provide food security to the tribal society. Along with this, it also helps in the conservation of soil, improvement of soil fertility, reclamation of degraded lands, stabilization of watersheds, protection of biodiversity, meeting the fuel requirement, etc. In fact, forest play effective role in the best utilization of the natural resources for sustainable crop production and socio-economic upliftment of the tribal community (Malik *et al.*, 2010).

Methodology

Since time immemorial, there is intimate association between tribes and forest foods, but still systematic and authentic information regarding forest foods of Jharkhand used by tribal people are very limited. Keeping these points in mind, a study was conducted in the tribal forest ecosystem to document such wild tribal foods with food security orientation.

Aim of the present study was to find out the role of forest foods in the life of tribes (living in or nearby forest ecosystem) to meet their basic needs, from foods to medicine. The study was conducted in Bokaro district of Jharkhand State in 2010 year.

Tools for the Data Collection

In initial phase, frequent visit were done for developing rapport with them, especially with the older people of the villages and well known youth persons who were linked with various government and non-government as well as other social activities of the villages.

A pre-structured schedule was prepared and it was pre-tested on non-sample group of tribal people. The necessary modification was made accordingly. Different types of tools for data collection were chosen as per the purpose of the research work. These tools were (i) Informal interview, (ii) Observation, (iii) Rapid Rural Appraisal (RRA) and (iv) Transect walk.

(i) *Informal Interview*: Through informal interview, data regarding income was obtained because tribal people extremely hesitated to provide any kind of information in this aspect. Therefore, accurate picture of their Socio-economic Status was very difficult to obtain. Only a rough estimation was done with the help of informal interview.

(ii) *Observation*: Tribal people were not able to say about "TYPE OF PLANT" whether it is herb, shrub or tree at the time of giving information about these forest foods. This type of data was collected by direct observation of the plants in the forest.

(iii) *Rapid Rural Appraisal (RRA)*: A rough list of forest based edible foods including roots, tubers, leafy and other vegetables, fruits, flowers and gum, etc., was obtained through RRA. Indirectly, the investigator tried to know their food consumption pattern or their dietary habits.

(iv) *Transact Walk*: After 2-3 informal visit to the village and conversation session with the different group of the village people, an informal team was made with active type of tribal men and women. The objective of the study was made very clear through deep conversation and discussion with them, after that the investigator went on "Transact Walk" along with them.

Data Collection Procedure

In the morning, when tribal women finished their first-half household chores and breakfast was done by all of their family members, the investigator entered the nearby forest along with them and returned before evening, so that those tribal women could finish their rest of the household chores as well as from the security point of view, since the forest was dense. During the visit, all the information given by them was noted seriously with great attention. This procedure was followed repeatedly until whole area of the forest was not covered.

After covering the full area of the forest, the schedule was filled randomly by informal interviewing a number of tribal people who had good knowledge about different types of the forest produce through regular visit of the forest, since a very long time, perhaps from their own childhood. Because in the tribal culture, in general tribal women involved in the collection work along with their children and spend a lot of time in the forest in gathering work. Tribal women and their children spend more time as compared to their male partners.

Method of Collection

Most of the forest tree are very high and the produce of these plants are plucked with the help of sticks. For collection, the tribal women go inside the forests and cover long distances ranging from 2-5 km. The collection season is spread over the whole year for different items, so these forest produce collection activities provide employment to the local tribal almost throughout the year. A majority of forest produce is available during the month of April to July. From an early age, tribal children become habituated for this collection work. Therefore, the literacy percentages of tribal children are recorded very less (Rout, 2010).

Observation, Result and Discussion

In Bokaro district, to have a rough estimation of the tribal's socio-economic status living in or nearby forest ecosystem, only 50 households were informally interviewed.

During the survey work, a lot of valuable information was obtained both from young tribal people who played active role in the collection of all relevant information needed for the study and from old people who had deep knowledge regarding forest produce and its multi-purpose use in their life.

(i) *Type of the Family*: In the study area, majority of the family, i.e., 94 per cent of the family were of nuclear type, whereas only 6 per cent families were of joint type. This information clears the picture of "shifting in the trend of traditional values in tribal society."

(ii) *Family Size*: Family size is the most important criteria to study the socio-economic status of any family. In the study area, majority of the family, i.e., 40 per cent of the family had total 4 numbers of the family members, whereas the second large group of the family, i.e., 32 per cent, had total 5 numbers of the family members. Only 2 per cent of the family had maximum total 8 family members. It is very good sign, from population control point of view.

(iii) *Family Income of the Households*: Since tribal people, whether men or women extremely hesitated to provide any kind of information regarding their income, so it was very difficult for the investigator to prepare data about their socio-economic status. The range of the Family Income through farming was Rs. 2,000 to Rs. 25,000 in a year. Tribal people were not very much clear about their earnings through agriculture. On the basis of the information provided by them, the range of Per Capita Income calculated was Rs. 500 to above Rs. 2,500.

(iv) *Education Status of Tribal Men and Women*: Education is the best instrument for the development of any society. Regarding educational status of the Head of the Family (HoF), majority of them, i.e., 60 per cent were educated up to the tenth class whereas second large group, i.e., 20 per cent of the HoF were illiterate. Whereas, only 12 per cent of the HoF was Inter and none was Graduate. It is very unfortunate condition, in spite of various Literacy programmes launched by the Jharkhand Government, quite large number of the people were still illiterate.

It is the female education which can bring revolution in the State, so it is very necessary to know the educational status of the Home-makers of the Jharkhand State. Regarding educational status of the tribal women, the majority, i.e., 58 per cent of the tribal women were illiterate whereas, the second large group, i.e., 30 per cent had education unto the sixth to the tenth class. Hardly 2 per cent of the tribal women were Intermediate pass. But, none was graduate.

The overall literacy rate among the STs has increased from 27.5 per cent at 1991 Census to 40.7 per cent at 2001 Census. Despite this improvement, the literacy rate among the tribes is much below in comparison to that of all STs at the national level (47.1%). It is very clear from this data that for bringing desired level of the development in the tribal society, they have to go a long way (www.jharkhand.nic.in).

Dietary Pattern of Tribals

During the study, a number of forest plants, which have great significance in their daily life from foods, medicine and livelihood need, were enumerated. All these plants are discussed here according to the different groups of the foods.

General Meal Pattern of Tribals

Their General Meal Pattern consists of three meal in a day throughout the year except in time of celebrations. They took breakfast in the morning because most of them had to go on their work either in the field or on labour work. In the noon period, they took lunch, which was the same rice and vegetable, cooked in the early morning. Tea was not a habit in their daily routine. Snacks was generally not taken by the adult persons of the family, it was given only to the children.

Their breakfast or lunch, both commonly consist of rice and vegetable; sometimes they took *dal* also. The most important thing is that they were very fond of non-vegetarian diet. Even in the prevalence of high poverty, they made some arrangement of it in any way, e.g., they caught small fish from river. *Chingri* fish was their most favourite fish.

Their diet comprises variety of unconventional and wild foods, viz., edible forms of flowers, fruits, tubers, leaves, stems, seeds, and wild mushrooms. Tribal mostly eat vegetables of leafy varieties, and depend on such natural products for their food. In the time of scarcity or when the staple food is in short supply, people are in general dependent upon various species of wild plants.

Forest as a Food Security Source for Jharkhand Tribal Population

Food security in India has recently been more alarming than ever before. Increasing pressure on land, environmental degradation, more frequent occurrence of natural calamities like flood, drought, etc., and resource differential between rich and poor have worsened the scenario. In newly formed, Jharkhand State, the situation is even more serious. Despite the availability of huge natural resources, more than 50 per cent of the total population lives below poverty line. According to National Sample Survey II (conducted in 2007-08) about 2 per cent of population suffers from acute and chronic hunger and 10 per cent from seasonal food insecurity. Settled agriculture does not provide them sufficient food and livelihood security. As most of the tribal farmers are small farmers having small and uneconomic land-holdings. They are able to grow foodgrains only for 8-10 months. For the other food items, e.g., fruits and vegetables, they generally depend on forest foods throughout the year. For the need of staple food (in remaining period), they depend on the forest. Edible forest produce provide essential nutrients and have been reported to improve both the physical and mental well-being of the tribal people (Sinha and Lakra, 2005).

Most Commonly Used Forest Foods in Jharkhand

(i) *Roots and Tubers*: Some of the most commonly consumed wild root and tuber vegetable were Ban-ole (*Amorphophallus bulbifer*), Colocassia (*Colocassia* spp.) and Ban-kundri (*Melothria heterophylla*). They are rich in carbohydrate and can be an important source of energy in our diets.

Processing and Preservation of Wild Roots and Tubers

In case of *Ban-ole*, its leaf, stem and tuber are eaten as vegetable. During rainy season, its leaves, petioles, and stem were ground with raw rice and *pitha* (small almond size product made up of tough rice paste) was made. Tuber was preserved by Oraon tribe by making *Bari* (small porous and light weight round balls flattened from below). For this purpose, they mashed the steamed tuber with tamarind juice and dry in the sun for use in the off-season.

During the crucial months, the tribal people collected tubers to supplement their meagre food available at home. Tubers of certain species were eaten as curries (e.g., *Ban-ole*). Some tubers were consumed by boiling, while some were cut, dried and made into flour and prepared their own kind of recipes. Most popular tubers were *Ban-ole* and *Ban-Kundri*.

(ii) *Green Leafy Vegetables*: Tribal people consumed a number of wild leaves in their diet round the year. Various types of wild leaves included in their dietary habit. The commonly consumed wild leaves eaten by them, were colocassia leaves (*Colocassia* sp.), Kana sag (*Commelina benghalensis*), Munga sag (*Moringa oleifera*), Khapra sag (*Trianthema monogyna*), Koinar sag (*Bauhinia purpurea*), Beng sag (*Hydrocotyle asiatica*), Neem sag (*Azadirachta indica*) and Putkal sag (*Ficus glabella*) etc.

Nutrient Composition of Wild Leafy Vegetables

The green leafy vegetables are rich source of calcium, iron, beta-carotene, vitamin-C, and folic acid (Gopalan, 2004). These leaves were very inexpensive and easily available in the local village market. They may contain all important nutrients, as per their food group, required for growth and maintenance of health. According to the NFHS III (conducted in 2005-06 in Jharkhand), approximate 80.5 per cent of the children (6-35 months age),73.4 per cent reproductive women (15-49 years age) and 68.2 per cent pregnant women in rural areas were found anemic (www.nfhsindia.org). Hence, wild leaves must be consumed by children, pregnant and nursing women to obtain much needed nutrients, i.e., beta-carotene, calcium and iron. This is particularly so important for predominantly Indian cereal-based diets of the poor who suffer from the dietary deficiency of these nutrients.

Processing and Preservation of Wild Leaves

Their diet were composed of various types of wild leaves, as per their availability in their respective season. They cooked fresh leafy vegetable along with rice or sun dried leafy powder is mixed in the rice gruel and consumed. They kept sun-dried leaf powder for off-season use or preserve various types of wild leaves by sun drying round the year.

(iii) *Other Vegetables*: The most commonly eaten other vegetables were Ban-Kundri (*Melothria heterophylla*), Karil/Sandana (*Dendrocalamus strictus*) and Ban-Karaila. In other vegetables Ban-kacchu (*Colocassia* sp.), wild Mushroom and Rugra were very popular among them. Vegetable which do not come under Green Leafy Vegetable and Roots and Tuber are classified under this head. Other vegetables not only add variety to the diet, but also provide other essential nutrients, e.g., Vitamin C and B complex. These vegetables are also the source of dietary fibre in the diet and provide bulk to the diet. (Gopalan, 2004)

Processing and Preservation of Other Wild Vegetables

Among all tribes, Karil/sandhana (*Dendrocalamus strictus*) is very famous for use in the lean period. *Karil* is new and young shoot of bamboo (Singh, 2007). It contains poisonous "Hydrocynic Acid" which they remove by boiling in water for half an hour repeatedly 3 to 4 times. They made good curries and also very tasty pickles. Surplus shoots were preserved by sun drying.

(iv) *Fruits*: The most commonly used wild fruits were Amada (*Spondias mangifera*), Ber (*Zizyphus jujuba*), Bhelwa (*Semecarpus anacardium*), Ban-Kundri, Kusum (*Schleichera trijuga*), Kend (*Diospyros melanoxylon*), Mahua (*Madhuca latifolia*) and Sakhua (*Shorea robusta*), etc.

Nutrient Composition of Wild Fruits

Fruits are good sources of beta-carotene and Vitamin C. Amada (*Spondias mangifera*) is rich sources of Vitamin C. Yellow fruits, e.g., Kend (*Diospyros melanoxylon*), Bhelwa (*Semecarpus anacardium*) and Kusum (*Schleichera trijuga*), etc., may have good amount of beta-carotene, a precursor of Vitamin A. The most widely used fruits, Sal (*Shorea robusta*) and Mahua (*Madhuca latifolia*) may have good source of carbohydrates and hence energy. Piyar (*Buchanania latifolia*) are good source of protein as it belongs to Dry Fruit group. Fruits contain dietary fiber-pectins which provide bulk to the diet and helps bowel movement.

Processing and Preservation of Wild Fruits

A large number of wild fruits were consumed by tribals. Some are eaten as vegetable when unripe, while ripen fruits were eaten as such. Unripe fruits were consumed as *bhujia, sabji* or as curries. Some of the fruits are pickled, e.g., Amada (*Spondias mangifera*) and some are made into *chutney*. Different fruits are eaten in different seasons.

(v) *Fat and Oils*: Visible fat and oil being a very expensive food item, the diets of the poor tribal people contain almost negligible or very little amount of visible fat. So, to meet their fat and oil need, newer sources of edible oil are being explored. In this aspect, tribal people knowledge will be of great significance, because there were a number of forest plants seeds from where good quality of edible vegetable oil was obtained, e.g., Mahua (*Madhuca latifolia*) Seed oil and Kusum (*Diospyros melanoxylon*) seed oil, etc.

Nutrient Composition of Wild Source of Visible Fat and Oil

All fats and oils are concentrated source of energy providing 9 kcal/g. In general, vegetable oils contain polyunsaturated fatty acids like linoleic and linolenic acids which are needed to meet the Essential Fatty Acids requirement. High fat intake is harmful leading to cardio-vascular complications. So, total calories from visible fat should not exceed 20 per cent, but to meet the minimum needs, 5-10 per cent of calories should be derived from visible fat and oil.

Miscellaneous

(vi) *Flowers*: Flowers of several forest plant were also cooked and eaten, e.g. Mahua flowers (*Madhuca latifolia*), Koraiya flowers (*Holarrhena anti dysenterica*) and Jirhul flowers (*Indigofera pulchella*) etc. Mahua and Putkal flowers (*Ficus glabella*) are available in summer season, while Jirhul flowers are available in spring seasons. These were very popular among tribal people.

(vii) *Wild Seeds*: Various types of wild seeds were also consumed by tribal. Some of them had very good market value. Seeds of Piyar (*Buchanania latifolia*) are highly expansive in present day market having Rs. 400/kg and considered very nutritious as they belongs to DRY FRUIT group.

(viii) *Wild Gum*: The gum of some forest plant is eaten by specific group of tribal. Asan (*Terminalia tomentosa*) is basically a timber tree but its gum is edible.

Processing and Preservation of Wild Flowers, Seeds and Gum

Mahua Flowers (*Madhuca latifolia*): The most popular flowers of the Jharkhand State is Mahua flowers. The cream coloured fleshy thalamus/corolla of Mahua flowers is sweet and used for various purposes and eaten in various forms. It is collected from the forest and sun-dried and kept for lean period use. They boiled flowers and then fried by adding salt and spices. During food scarcity, dried Mahua flowers were boiled either with tamarind seeds or Sakhua/Sal seeds and form an important preparation called "*Lattha*" (sweet balls made up of roasted wheat flours or maize flours and Mahua flowers) of tribal diet. Sakhua/Sal (*Shorea robusta*) and tamarind seeds were popular among all tribes and widely eaten. These seeds were mixed with dried Mahua flowers (*Madhuca latifolia*), boiled and consumed. It is very important foodstuff for the poor tribal group. Mahua seeds yield very famous edible oil, which was used as a substitute for *vanaspati*. Seeds of tender and immature fruits of Ber (*Zizyphus jujuba*), Kend (*Diospyros melanoxylon*) and Kusum (*Schleichera trijuga*) were eaten.

Asan is basically a timber tree but its gum is edible. The gum of Asan (*Terminalia tomentosa*) is roasted in the direct fire, which swells like papad and then eaten.

Medicinal Properties of the Forest Produce for Health Security

There were good number of old very knowledgeable Tribal people who had immense information regarding the medicinal properties of

Asan (*Terminalia tomentosa*)

the various forest plants, which have great significance for their health security. Some of the information are given below.

Asan (*Terminalia tomentosa*)**:** The bark of the *Asan* is diuretic and cardio-tonic. It is also used in the treatment of the diarrhoea and wounds.

Bhelwa (*Semecarpus anacardium*)**:** It is very helpful in the cough and gas problem and also used in the complications related with nervous system. It is best medicine in case of skin diseases. Bhelwa seed oil is extremely useful in the rheumatism and leprosy treatment. Ashes of this plant is very effective in the snake-bite and scorpion-sting. But, it should be used with great precautions and under well experienced and knowledgeable person or learned man.

Beng-sag (*Hydrocotyle asiatica*): Beng-sag is useful in the treatment of various complications like digestion disorders and blood related diseases. Tonic made from the leaves of the Beng-sag improves memory. It is very effective in case of skin, leprosy and nerve related disorders. Beng-sag is also helpful in improving eye-sight.

Ban-Bhindi (*Hibiscus abelmoschus*): This plant, particularly its leaves and seeds, looks very similar to ladies finger plant. *Ban-Bhindi* seed, roots and leaves is used in the treatment of improper urination or in case of difficult *urine excretion.* Its seed oil is highly beneficial for heart patients and in the treatment of gas problem and in body

Ban-ole (*Amorphophallus bulbifer*)

Beng Sag (*Hydrocotyle asiatica*)

pain. The regular use of it improves the eye sight. The throat swelling and related voice problem is cured by inhaling the fumes of dried *Ban-Bhindi* plant.

Jirhul Flowers (*Indigofera pulchella*)

Karil (*Dendrocalamus strictus*)

Kusum (*Schleichera trijuga*): The bark of the *Kusum* tree and its oil is used to cure itching problem. The *Kusum* seed oil is efficient and stimulating agent for scalp, both for cleaning and promoting growth of the hair.

Kusum (*Schleichera trijuga*)

Kend (*Diospyros melanoxylon*)

Kend (*Diospyros melanoxylon*): The *Kend* seed oil is given in the diarrhoea and dysentery treatment. Fruits are used to gargle for treating the throat problem. Fruit juice is used for application on wound. Bark is also used to cure dysentery and fever.

Koraiya Flower (*Holarrhena antidysenterica*)

Mahua Flower (*Madhuca latifolia*)

Koraiya flowers (*Holarrhena antidysenterica*): It is very effective in malaria fever, joint pain as well as in controlling the loose motion. It acts as blood purifier. It is also given in the treatment of gout and piles. It is used in the kidney related disorders.

Sun Dried Mahua Flower (*Madhuca latifolia*)

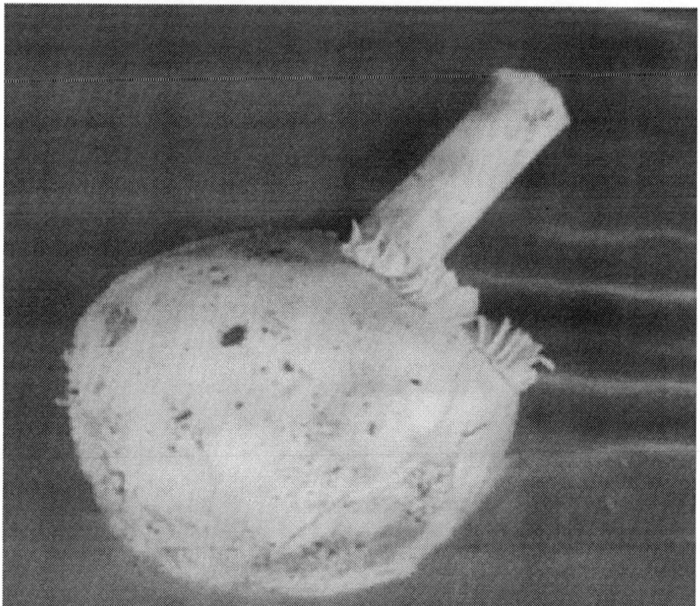

Wild Mushroom

Rugra

Jamun (*Syzygium cumini*): From *Jamun*, a number of indigenous health products are made, in which fruit vinegar (made up of *Jamun*) is very popular among diabetic patients. Its bark and seed is highly beneficial in the treatment of diabetes and any kind of digestion related problem. It is also helpful in the constipation treatment.

Mahua (*Madhuca latifolia*): The bark of the plant is a remedy for rheumatism and itching problem. From flowers, a distilled liquor is prepared which is astringent in taste and regarded as cool, appetizing and nutritious tonic used in cough. *Mahua* flowers are also used by frying in ghee and eaten by persons suffering from piles. It is also used in the case of fish poisoning. Its seed oil is excellent for skin disease and any kind of stomach pain or problem whereas its twig is very good for pyorrhoea. Excessive use of its seed oil is harmful, so it should be taken under well experienced person.

Munga (*Moringa oleifera*): *Munga* leaves, due to hypocholesterolemic effect, are very effective for blood pressure patients.

Sakhua (*Shorea robusta*)

Piyar (*Buchanania latifolia*): It is very famous dry fruit. Its kernel oil is used as substitute for almond in medicinal preparations. It is very effective in the removal of cough and constipation. The kernel of *Piyar* is used as ointment in skin disease.

Sakhua/Sal (*Shorea Robusta*): *Sakhua* is very good blood purifier. It is used in the treatment of worm infestation. Its bark and leaves is cooked in the sesame (*til*) oil and used to cure the ear pain. The resins of the *Sakhua* is used in the diarrhoea and dysentery.

Forest as a Livelihood Security Source

In India, approximately 50 million tribal people receive substantial proportions of their cash and in kind income from forest produce while about 200 to 300 million village people depend upon forest products in varying degrees. At the national level, over 50 per cent of forest revenue and about 70 per cent of forest export revenue comes from the

Tribal people as Volunteer

non-timber type forest produce, mostly unprocessed and raw. Small scale forest-based enterprises provide up to 50 per cent of the income for about 25 per cent of Indians rural labour force (Prasad, 2007).

Vast areas of Jharkhand harbour natural luxuriant forests of Sakhua (*Shorea robusta*) along with the other diverse associate species. *Kend* leaves (*bidi leaves*) and *Sakhua* leaves for "*Dona-Pattal*" (disposable plates and cup made of leaves) is the major source of revenue providing considerable employment to tribal during the lean period of agriculture. Most of the forest produce are available for particular season but they are utilized all the year round.

Forest as an Instrument of Women Empowerment

Some 50 million tribal people in India depend on forest produce for meeting their subsistence, consumption and income needs. A large proportion of employment generated by forest produce accrues to women. The scope of forest produces are so wide and fruitful that if properly managed then it could be very good source of tribal women empowerment and alleviate poverty (Prasad, 2007).

Marketing of the Forest Foods

Approximately 50 million tribal people receive a substantial proportion of their foods and income from these forest foods. In Jharkhand, the average highest income is obtained from the sale of Mahua flowers (*Madhuca latifolia*) having the price approximate Rs. 25 per kg. The second highest valuable item is Jamun fruit (*Syzygium cumini*) having the price approximate Rs. 30 per Kg. Wild Mushroom and Rugra, a type of edible fungi, are sold in the urban market having price of Rs. 160 per kg and Rs. 120 per kg respectively (*Note*—the scientific name of the wild Mushroom and Rugra is not known to the investigator during the research work). Jamun (*Syzygium cumini*), Mushroom and Rugra are available only for short period, i.e., 2-3 months. These are very popular forest foods not only for rural tribal people even among urban people also. Therefore, the tribal people come to the urban areas to sell their collected forest foods to get maximum return.

Marketing of most forest produce were carried out in traditional ways. Quality is not maintained as some of these products are perishable and season-oriented and also due to usage of unscientific methods of collection and storage. As a result, value addition of many products and more specifically in the case of pharmaceuticals is not done effectively.

Normally, forest produce were collected haphazardly and sold locally in the village market. Still, the demand of these forest foods was very high among both local and urban people due to its natural and unique taste, e.g., wild Mushroom and Rugra. Processing of forest produce into value-added products through simple technologies is very limited. So, research is needed for the improvement in the situation.

Constraints in the Utilization of Forest Produce

There are a number of factors which hinder the best utilization of the forest foods. These are as follows :

1. Forest produce have poor storage and preservation quality, therefore, deterioration is a common phenomenon. According to FAO (2002), most of the forest produce are collected in a particular season although they are utilized all the year round and as much as 15 per cent spoil during storage.
2. Circumstances surroundings the forest areas, warrant development of improved technology for storage and preservation of forest produce.
3. The most crucial problem is marketing of forest produce. The tribal collector at the lowest end receives a pitifully low price for the product he collects, while the consumer has to pay the inflated price, the middle man appropriating the large difference. Unless marketing problem is addressed properly it will not be possible to make forest produce a reliable source of income. A definite procurement strategy for forest produce is called for.
4. Most forest produce provide employment for a part of the year because processing of these forest produce is still poorly developed.
5. The potential of many forest produce is not being fully utilized due to insufficient knowledge and experience on appropriate processing techniques and development of the end products.
6. Extracting, processing and production of the most of the forest produce are carried out in traditional ways.
7. The other concept is closely related to the above—though commercial crops have support prices, the most perishable items like forest produces do not have any such support.

Suggestions for Best Utilization of Valuable Forest Produce

Value Addition of Forest Produce

First stage processing should be organised at local level for value addition and employment generation. Local processing includes grading, purifying, storing in congenial conditions should be done to reduce post-harvest losses and improving thereby the product quality to fetch higher price. Those simple and low cost processing technologies that are appropriate to forest produce have to be developed and ensuring better value for the material collected by them. This will result in poverty alleviation.

Urgent Directions for Best Accruing from Forest Produce

During the research work among tribal people, it was highly noticed by the investigator that just due to the ignorance or lack of awareness regarding various simple but important things, e.g., "Basic concept of Weight and Measures" many times, tribal people could not be able to earn their genuine income. Therefore, the following urgent directions are needed for the best accruing from forest produce :

(i) Community Based Awareness Programme on the "*Basic Concept of Weight, Measures and Account Keeping*": Since forest medicinal plants and foods produce are very costly as well as highly perishable by nature, e.g., Ban-Bhindi, Wild Mushroom and Rugra etc. These are also available only in the particular season like Jirhul flowers (*Indigofera pulchella*) in the spring season and Koraiya flowers (*Holarrhena antidysenterica*) in the summer season, etc. It is very important for tribal people, especially for tribal women among which the literacy rate is very low, they must have the basic concept of, Weight, Measures and Account Keeping.

Some arrangement should be made by the Jharkhand Government for conducting "Community Based Awareness Programme" for imparting to them the basic concept of Weight, Measures and Account Keeping on urgent basis.

(ii) *Simple and Cost-Effective Preservation Technology*: During the research work, it was very clear that Sun Drying or Natural Dehydration was the most widely used method of preservation by the tribal people, e.g., sun-dried Mahua flowers, sun-dried young and new shoot of Bamboo or Karil, sun-dried powder of various wild leaves, etc., were very popular among them.

Since sunlight is available in plenty round the year, therefore, some arrangement should be made to provide them "Solar Driers" from Jharkhand Government for timely and fast drying of these valuable forest produce. The soft and delicate forest produce will be also saved from the effect of overdrying or undesirable drying.

In this aspect, the technology, Zero Energy Cool Chamber (ZECC) and Solar Driers (SD) can be provided in the nearby areas to preserve highly perishable forest produce like wild Mushrooms, Rugra, Jirhul flowers, Munga flowers, Koraiya flowers, etc., With the help of these two simple and cost-effective technologies, the preservation of important forest produce will be done on time, which leads to enhanced shelf life of these produce. From both ways, they will also be able to check their spoilage.

Normally sunlight is available for long duration, so solar cooker will be very helpful in cooking of these forest produce. Solar cooker will also reduce the dependency of tribal people on forest fuel, which will facilitate in conservation of forest resources. In other words, Solar Cooker will act as alternative source of fuel for the tribal people. Therefore, these should be made available from the Jharkhand Government side to them at subsidised rate.

(iii) *Adequate Storage Practices*: According to FAO (2002), approximately 15 per cent of the forest produce gets spoiled. The tribal people storing practices were not adequate and sufficient by nature to store their forest produce for desired period. With the help of Zero Energy Cool Chamber (ZECC), the shelf life of highly perishable forest produce, e.g., wild flowers (Jirhul flowers, Munga flowers), Wild Mushroom and Rugra, etc., will be enhanced, the spoilage of these produce will reduce and ultimately they will get more income in return. The infrastructure of Zero Energy Cool Chamber should be provided from the Jharkhand Government to them.

(iv) *Packaging Practices for the Forest Produce*: In general, tribal people had no idea of simple packaging practices. They sold their forest produce in traditional ways. Under primary processing, tribal people should be made aware about "Simple Packaging" practices. Here, also the basic Community Awareness Programme should be launched to make them aware regarding modern packaging practices and marketing system of today's world.

Tribal people live in harmony with nature. Since time immemorial, forest has played a vital role in providing basic needs in the form of food, medicine and livelihood source. The diversity of these products is so large and their utility in their day-to-day life is so great that it becomes essential to incorporate these forest foods into the future human security programmes.

REFERENCES

FAO. 2002. *Non-Wood Forest Products in 15 Countries of Tropical Asia: An Overview*. P. Vantomme, A. Markkula and R.N. Leslie (eds.), Bangkok.

Gopalan, C.; B.V. Ramasastri and Balasubramanium, S.C. 2004. *Nutritive Value of Indian Foods*. Hyderabad, National Institute of Nutrition. ICMR. 29-30.

Jharkhand Forest Resources Survey. 2003. *Division Survey Report on Availability and Resources of Medicinal Plants of North and South Chotanagpur Divisions*. Forest Department, 68-88.

Lipper, L. 2005. Forest degradation and food security. In *FAO Corporate Document Repository* cited on x7273e05.htm dated 2-9-2010.

Malik, M.S.; P. Kaushal and R.B. Sah. 2010. *Socio-economic Upliftment of Tribal Communities in Jharkhand through Agroforestry based Farming System*. Birsa Agricultural University, Ranchi. Cited on http://jharenvis.nic.in on 26-7-2010.

Prasad, I. 2007. *Quantitative analysis of NTFPs in Mahilong and Horhap Forest Beats of Ranchi East Forest Division and their Sustained Management*. Thesis, M.Sc. Birsa Agricultural University, Ranchi.

Rout, S.D.; S.K. Panda; N. Mishra and T. Panda. 2010. "Role of tribals in collection of commercial non-timber forest products in Mayurbhanj district, Orissa." *Stud Tribes Tribals*, 8(1): 21-25.

Singh, R.K.; A. Singh and A.K. Sureja. 2007. "Sustainable Use of Ethnobotanical Resources." *Indian Journal of Traditional Knowledge*, 6(3): 521-530.

Sinha, R. and V. Lakra. 2005. "Wild tribal food plants of Orissa." *Indian Journal of Traditional Knowledge* 4(3): 246-252.

Sinha, R. and V. Lakra. 2007. "Edible weeds of tribals of Jharkhand, Orissa and West Bengal." *Indian Journal of Traditional Knowledge*, 6(1): 217-222.

www.jharkhand.nic.in

www.nfhsindia.org

9

Community Based Ecotourism at Jhilmil Jheel Conservation Reserve, Uttarakhand : Identification of Potentials

Bipin Kumar, Dinesh Bhatt, Bitapi C. Sinha and *Vir Singh*

Introduction

Tourism is the world's largest industry. It accounts for more than 10 per cent of the total employment, 11 per cent of global GDP, and total tourist trips are predicted to increase to 1.6 billion by 2020. As such it has a major and increasing impact on both people and nature. Over the past 15 years, ecotourism has become one of the fastest growing sectors of the tourism industry, growing three times faster than the industry as a whole. At the same time, ecotourism is being increasingly viewed by local and indigenous communities as an important tool for promoting sustainable livelihoods, cultural preservation, and biodiversity conservation (Honey, 2006). The International Ecotourism Society (TIES, 2006) defines ecotourism as "responsible travel to natural areas that conserves the environment and improves the well-being of local people". Ecotourism has, in essence, three core tenants. It holds that, done well, it can (1) protect and enhance the environment; (2) respect local cultures and provide tangible benefits to host communities; and (3) be educational as well as enjoyable for the travellers (Honey, 2006).

Ecotourism is a frequently debated term and it emerged in the late 1980s as a direct result of the world's acknowledgment reaction to sustainable practice and global ecological practice (Diamantis, 2000). Ceballos-Lascurain (1987, cited in Boo 1990) has defined it as follows: "Travelling to relatively undisturbed or uncontaminated natural areas with the specific objective of studying, admiring, and enjoying the scenery and its wild plants and animals, as well as any existing cultural manifestations (both past and present) found in these areas". The International Ecotourism Society also defines ecotourism as responsible travel to natural areas that conserves the environment and sustains the well-being of local people. This definition not only implies that there should be a recognition of, and positive support for, the conservation of natural resources, both by suppliers and consumers, but also that there is a necessary social dimension to ecotourism. The increased environmental awareness among visitors of environmental protection and the shifting demand from mass tourism to new form of tourism significantly contributes to the development of ecotourism particularly in natural protected areas. The positive linkage between conservation and ecotourism is the key factor attracting developing nations to promote this new form of tourism to meet current tourist demand as well as to conserve natural attractions in their areas. It is viewed as a strategy for developing countries to create sustainable economic development while pursuing conservation objectives by balancing between the often conflicting goals of poverty reduction and biodiversity conservation (Cusack and Dixon, 2006). With this reason, ecotourism is considered as win-win solution to advance resource conservation and poverty reduction, which is mutually reinforcing goals if it is well-planned and managed by the local communities. Increasingly, this general sector of the market is also called 'nature tourism'. True 'ecotourism', however, requires a proactive approach that seeks to mitigate the negative and enhance the positive impacts of nature tourism. The term "*community-based ecotourism* (*CBET*)" should be used for those ventures based on the high degree of community control (hence, demand for large proportion of benefits) rather than those almost controlled by outside operators (Jones, 2005). Community-based ecotourism takes this social dimension a stage further. This is a form of ecotourism where the local community has substantial control over, and involvement in its development and management, and a major proportion of the benefits remain within the community. How the community is defined will depend on the social and institutional structures in the area concerned, but the definition implies some kind of collective responsibility and approval by representative bodies. In many places, particularly those inhabited by indigenous peoples, there are collective rights over lands and resources.

However, it must also embrace individual initiatives within the community. Some further general characteristics of ecotourism have been identified by UNEP and the World Tourism Organization as:

- involving appreciation not only of nature, but also of indigenous cultures prevailing in;
- natural areas, as part of the visitor experience;
- containing education and interpretation as part of the tourist offer;
- generally, but not exclusively, organized for small groups by small, specialized man;
- locally owned businesses (while recognizing that foreign operators also market and operate ecotourism);
- minimizing negative impacts on the natural and socio-cultural environment;
- supporting the protection of natural areas by generating economic benefits for the managers of natural areas;
- providing alternative income and employment for local communities; and
- it generates much needed foreign currency, both locally and nationally and increasing local and visitor awareness of conservation.

Location of the Jhilmil Jheel

Jhilmil Jheel is situated between N 29° 32′ to 29° 50′ and E 78° to 78° 15′ covering an area of 3,800 ha of Reserve Forest. The altitude of the area varies from 200 to 250 metres above msl. It is located between the Haridwar-Najibabad highway and the natural course of the Ganga to the south of it in Chidiyapur Range of Haridwar forest division, Uttarakhand.

Geology, Drainage and Climate

It is located at the junction of the Bhabhar and Terai formations extending up to the Ganga River. The texture of the soil varies from fine sand to clayey loam. The central part of study area is a wetland and part of the area gets submerged during the monsoon months. A number of small rivulets emerge from the woodland and discharge into Jheel, which finally drain into the Ganga River. It receives water from the northerly formations of the Shivaliks of Chidiyapur range or even further beyond as underground streams or 'Choyas' as well as from the flood of River Ganga. The area experiences sub-tropical climate, coldest month being January when temperature goes as low as 2°C have been recorded. Winter rains and subsequent heavy fog are also prevalent in the area. Summer is very hot and humid like any other *Terai* area when temperature soars up to 44°C, often accompanied by hot dust storms. The area experiences monsoon by the end of June and rains continue till September. Average rainfall in the area is 1800 mm per annum and most of the rainfalls between July to September.

Land-use Pattern

Jhilmil Jheel falls within Chidiyapur range of Haridwar forest division as shown in Map 9.1. The study area includes Comptt: Nos Nalowala 3, 4, 5, 6, 7a, 7b, 8a, 8b, 9a, 9b, 10a, 10b, Amsot 1a, 1b, Ganga 1a and 1b. The Jheel is flanked by Tantpur village to the west having about 130 farmer families living within an area of 303.265 ha. The adjoining area of Jhilmil Jheel lies in the flood plain of River Ganga. The villagers cultivate largely sugarcane, wheat and paddy. In addition, some Gujjar families live around the Jheel and use the surrounds of the Jheel for grazing.

Floral Diversity

The preliminary effort during the first sighting of Swamp deer in this area shows the dominance of *Typha angustifolia* and *Phragmities karka* in the main swamp of Jhilmil Jheel, typical of upper Gangetic flood plain habitat. A few grassy galdes dominated by *Cynodon dactylon* and *Imperata cylindrica* form the regular grazing ground for domestic livestock. The margins of swamp have mixed woodland characterized by miscellaneous tree species especially *Ficus benghalensis, Azadirachta indica, Anogeissus latifolia, Streblus asper, Adina cordifolia, Acacia catechu, Dalbergia sissoo* and *Zizyphus mauritiana.* The other floral components of this area are *Cordia obliqua, Asparagus adscendens, Adhatoda zeylanica, Cassia fistula, Aerva lanata, Ehretia laevis, Capparis sepiaria, Ipomoea carnea, Dendrocalamus strictus, Limonia acidissima, Sida cordiflolia, Grewia tiliifolia, Schleichera oleosa, Bauhinia racemosa, Aegle marmelos, Murraya koenighii,*

Helecteris isora, Mitragyna parviflora, Chrysopogon fulvus, Carissa carandas, Hemigraphis latibrosa, Adiantum edgeworthii, etc. Among aquatic and semi-aquatic plants are *Ammania baccifera, Medicago polymorpha, Phyla nodiflora, Ranunculus scleratus, Typha angustifolia, Cyperus alopecuroides, Veronica anagalis-aquatica, Polygonum barbatum* and *Ceratophyllum demersum.*

Faunal Diversity

Among the endangered fauna inhabiting the area are Swamp deer (*Cervus duvauceli*), only recently confirmed authentically in Uttarakhand. A striking feature of this area is the presence of five species of deer in this area. The shallow water with swampy habitat also attracts a variety of water fowl and more than 30 species of both migratory and resident birds have been recorded. Raptors such as Marsh harrier, Egyptian eagle, Honey buzzard and Osprey have been sighted around Jhilmil Jheel. As per information gathered from local inhabitants a group of 4-6 elephants and good number of nilgai regularly raid the sugarcane and paddy crops. The forest edge around Jhilmil Jheel is also extremely rich in herpetofauna, especially Python and entomofauna as also the Phragmites patches. Another striking feature of the area is presence of very high concentration of honey bees and high production of honey.

It was interesting to see large flocks of Braminy ducks in the cultivation around Jhilmil Jheel without any fear of the presence of people around. The behaviour of the birds shows that they feel safe here, possibly because of the conservation friendly attitude of the local people towards wild animals.

Data Source and Collection

The present study utilizes various data sources and data collection methods to analyze the potentials of community-based ecotourism at Jhilmil Jheel conservation reserves. These included structured questionnaire and attitude survey of local people and SWOT analysis of different stakeholders.

Survey of the Jhilmil Jheel Conservation Area

To find out the current status of the community around Jhilmil Jheel a structured questionnaire survey was conducted. The whole village

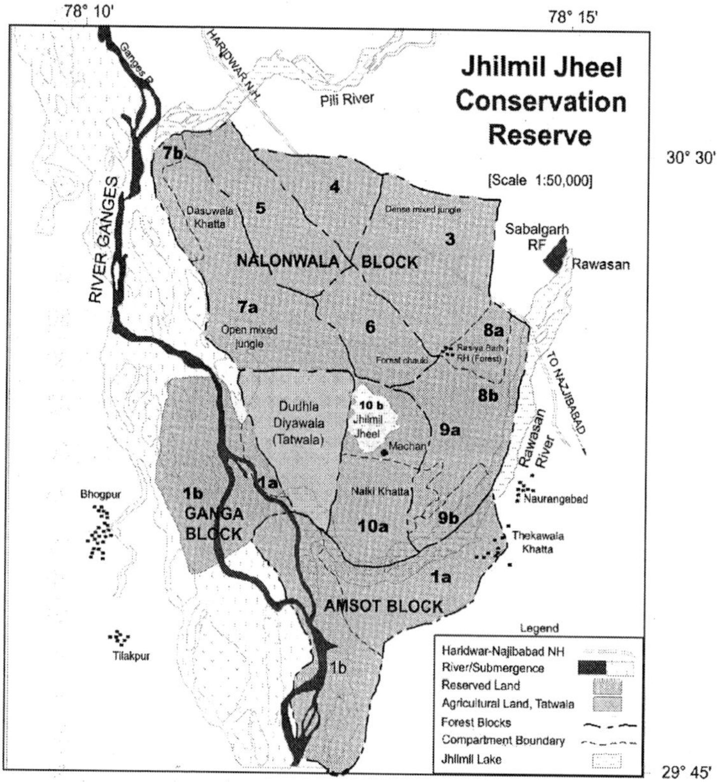

Fig. 9.1: Jhilmil Jheel Conservation Reserve

of Tantwala [respondent (N) =131] was covered through the survey. Out of 131, there were 24 Gujjars and rest 107 villagers (head of family) were interviewed. The questionnaire included both fixed response and open ended questions.

Attitude Survey

To identify the attitude about the conservation of Jheel, tourism, and extra income by the tourism and contribution in conservation of forest and wetlands (Jhilmil Jheel) of local people, an attitude survey was conducted randomly at the Tantwala village.

The main objectives of this survey were that to find out the attitude of local people about the Jhilmil Jheel and nearby forest. 113 local people interviewed by this survey. Out of 113, there were 14 Gujjars and 99 villagers. Respondents answered each attitude statement according to their strength of agreement by the following attitude level scores:

> 1=strongly agree, 2=agree, 3=don't know, 4=disagree, 5=strongly disagree.

Results and Discussion

Stakeholder Analysis

A stakeholder is any person or group or organization, i.e., involved or interested or impacted by "what is being done". Stakeholders can be defined as any person, group or organization that are affected by or can affect decisions concerning the project or those who passes a direct, significant and specific stake in the project.

Thus identifying stakeholders and creating a sense of partnership among them is a crucial task towards successful CBET development at the Jhilmil Jheel area, these are following (Table 9.1).

Table 9.1: Stakeholders and their stake at Jhilmil Jheel.

Stakeholders	*Stake*
Villagers	Homestead, Guide, Vehicle drivers
Agriculturist	Homestead, Guide, Vehicle drivers
Gujjars	Guides
Forest Department	Protection, conservation and revenue generation
Tourism Department	Revenue generation
Institutions	Training and education

SWOT Analysis

To develop corporate missions, objectives, strategies and policies, various strategic formulations are done for long-range planning processes. It begins with the situational analysis. For making strategic decisions, strategic factors in the light of current situation are analyzed using SWOT, an acronym for Strengths, Weaknesses, Opportunities and Threats (Wheelen *et al.*, 2004). Hence, here an attempt has been made to identify Strengths, Weaknesses, Opportunities and Threats of community-based ecotourism. The results are presented in (Table 9.2).

Table 9.2: SWOT analysis of stakeholders

Stakeholders	*Strength*	*Weakness*	*Opportunities*	*Threats*
Local Communities (Villagers & Gujjars)	*Knowledgeable about area's biodiversity and culture. Sustainable and accountable in the long run.*	*Lack of experiences. Capital and the reach to develop and market products.*	*Growing interest in cultural heritage and rural lifestyles. Visitors want to have first hand interaction with local people.*	*May lead to disruption to society and inequalities. Environmental impacts and damage to resources.*
Forest Department	*Custodian of natural resources. Professionally competent and committed staff. Strong interface with local communities. Scope for innovation.*	*Restrictive rules and regulation. Inadequate funding for eco-tourism. Weak personnel management. Limited knowledge and skills of eco-tourism. Inadequate infrastructure.*	*Expanding demand for eco-tourism. Availability of external funding for projects. Availability of external support and organizations for capacity building.*	*Increase pressures on natural resources. Lack of regulatory monitoring mechanism.*
Institutions	Expertise in training and research. Capacity to undertake project and provide services.	Scarcity of human and financial resources. Lack of practical experience in implementation of projects and running of eco-tourism enterprises.	Demand from Government and private agencies. Partnership with private sector/external agencies for funds.	Dependency on the host for implementation. Lack of govt. support and co-ordination between various organizations.
Tourism Department (not present at the moment)	Mandate and institutional framework for promotion and development of tourism. Authority to make policies and sanction funds.	Poor co-ordination with other enforcement agencies. Lack of financial autonomy in decision-making. Unskilled staff with low motivation	Stakeholder interest in benefit from ecotourism. Global and national Awareness on green issues. Entrepreneurial potential of Indians.	Unregulated growth of tourism and conflicts amongst stakeholders. Lack of adequate financial and human resources.

Existing Infrastructure

Results of questionnaire survey indicate that 56 per cent houses were *kutcha*, while 37 per cent houses were *pucca*, and 37 per cent houses were both *kutcha* and *pucca* in Tantwala village. *Pucca* houses were made of cement and bricks and *kutcha* houses were made of mud plastered wall with thatched roofs (Table 9.3).

Table 9.3: Infrastructure of Tantwala village

House	*Kutcha*	*Pucca*	*Both*
Villagers	60 (56%)	8 (37%)	40 (37%)
Gujjars	23 (100%)	—	—

Communities Attitude towards CBET

Ninety per cent local people strongly agree to the idea of the community-based ecotourism in Jhilmil Jheel area. They felt that through this they would be in a position to earn some additional money and it would be a motivation to protect the conservation reserve.

Willingness to contribute to CBET

Table 9.4 shows the willingness of the natives of Tantwala village to participate and contribute in activities to community-based tourism. 93 per cent of the respondents were keen primarily to work as guides along with other activities. The villagers are also ready to either spare a room from the existing infrastructure or build an additional room to provide accommodation to the visitors. 84 per cent of the respondents are willing to keep visitors as guests but 16 per cent respondents do not want to keep guests. The reasons noted were lack of infrastructure and social reasons. Since a large number of respondents agreed to keeping visitors in their houses, they were also asked if they would be willing to renovate their houses to suit the needs of tourism, 56 per cent villagers agreed and 44 per cent were against spending money for renovation of their houses initially (Table 9.4).

Expected Tariff for Accommodation and Other Facilities

To the question on how much would they like to charge for the services provided to tourists, 71 per cent villagers want to get rupees 50-200,

Table 9.4: Villagers willing to contribute for CBET

Activity	*No. of respondents willing to contribute*
Guide	122 (93%)
Food and Beverage	110 (84%)
Homestead	110 (84%)
Vehicle	35 (27%)
Others	72 (55%)

10 per cent villagers want to get rupees 200-400, 10 per cent villagers want to get rupees 400-500 and only one villager want to get more than rupees 500 for a day. About 8 per cent villagers do not expect any fee from tourists.

Table 9.5: Tariff expected by villagers from tourists.

Tariff (Rs.)	*Number of respondents*
50-200	93 (78%)
200-400	13 (11%)
400-500	13 (11%)
>500	1
No Fee	11 (8%)

Awareness about the Conservation Reserve

Cent per cent respondents are aware of Jhilmil Jheel as a conservation reserve. Results from attitude survey show that 18 per cent villagers strongly agree, 65 per cent agree, 6 per cent do not know, 8 per cent disagree and 3 per cent strongly disagree that the use of forest and wetland resources and their conservation should be done together. 2 per cent respondent agree, 6 per cent disagree that it is the waste of money and resources to conserve wetlands and surrounding forests when people are poor and are short of land and 73 per cent local people do not know about this statement (Table 9.6).

Conclusion

Jhilmil Jheel conservation reserve currently does not have any tourism but according to the study it has more potentials of community-based ecotourism. All villagers of Tantwala are aware about conservation reserve and they want to contribute in some way to the conservation of forest and Jhilmil Jheel because they know that to conserve different

Table 9.6: Results from attitude survey

Statement	*Strongly Agree*	*Agree*	*Don't know*	*Disagree*	*Strongly Disagree*
To conserve different kinds of wildlife, conservation of Jheel is also important	80%	20%	—	—	—
Jheel can be a means of entertainment and tourism	90%	10%	—	—	—
Tourism can bring extra income for the people in this region	73%	22%	2%	3%	—
It is waste of money and resources to conserve wetland and surrounding forest when people are poor and are short of land	—	21%	73%	6%	—
The use of forest and wetland resources and their conservation should be done together	18%	65%	6%	8%	3%
You want to contribute in some way to the conservation of forest and wetland	52%	48%	—	—	—

kinds of wildlife, conservation of Jheel is also important and they also think that Jheel can be a means of entertainment and tourism which can be beneficial for the local people because tourism can bring extra income. In addition, it has the abundance of wildlife, cultural heritage and scenic beauty.

Recommendations

To develop community-based ecotourism at Jhilmil Jheel conservation reserve area, some suggestions are as following :

Ecotourism Planning

The different actors in the reserve should engage in a participatory planning process to develop a strategic public use plan for the reserve. The planning approach should be based on the limits of acceptable change planning methodology that would incorporate cultural and social parameters into the monitoring and management process. Throughout this planning process, community member in fact should be the primary stakeholders and they should be involved throughout this process as designers and facilitators so that the final product is owned by them.

Community-based Zoning and Regulations of Tourism Land Use

Each area planned for tourism attractions should zone these areas as public use and specifically link conservation goals to these zones through an interdisciplinary dialogue including ecologists, wildlife biologists, local community members and park managers. These zones could be selected for their ecological and tourism values. The zones should integrate specific restrictions that are acceptable to the local residents and provide a reasonable measure of protection for the long-term viability of the tourism attraction.

Marketing and Promotion

The issue of marketing is very sensitive and any attempt should be taken with great caution. No new measure in marketing efforts should take place without the comprehensive plan and limits of acceptable change monitoring system in place. Great care should be taken with large-scale marketing efforts to grow visitation or to "make tourism pay."

Site Specific Recommendations

(a) Monitor and strengthen the capacity of community groups and service providers in households and organizational money management through targeted workshops, supervision and accompaniment.
(b) Involve the community members in a participatory planning process to define their vision for the future of ecotourism, the development of nearby attractions and the marketing of the conservation reserve as the primary destination.
(c) Focus on monitoring wildlife species in terms of ecological, tourism and cultural importance. Currently there is the lack of information about wildlife population dynamics around Jhilmil Jheel. A long-term, technologically appropriate monitoring programme of two or three indicator species would help to understand the impact of tourism on wildlife populations.
(d) Incorporate, train and supervise individuals who currently use the area as guides. With their forest and wildlife knowledge,

they would be excellent candidates as guides and this work could provide a direct economic incentive for them to protect species of concern around the community (Neilsen, 2001).

Capacity Building

The most urgent training needs in community-based ecotourism are at the field level. The success or failure of ecotourism ventures, as well as the protection of vulnerable environmental resources, are largely dependent on the skills, knowledge, dedication, and practices of field guides. Therefore, particular attention should be focused on providing high quality training to guides. Identified participants should undergo a minimum of 15 days initial training after which they should be given lecture to work as to nature guides. Regular refresher courses should be organized during the off-season. Besides knowledge of flora and fauna the participants should also be made familiar in hospitality management. The participants should be made aware of what ecotourism is all about and how it is conducted so that they are able to provide reasonable inputs to the success of the initiative.

REFERENCES

Boo, E. 1990. *Ecotourism: The Potentials and the Pitfalls.* World Wildlife Fund, Washington.

Ceballos-Lascura´in, H. 1996. *Tourism, Ecotourism and Protected Areas: The State of Nature-based Tourism Around the World and Guidelines for its Development.* Gland, Switzerland, and Cambridge, UK: IUCN.

Cusak, D. and L. Dixon. 2006. "Community-based Ecotourism and Sustainability." *Journal of Sustainable Forestry*, 22: 157-182.

Diamantis, D. 2000. "Ecotourism and Sustainability in Mediterranean Islands." *Thunderbird International Business Review*, 42(4), 427-443.

Honey, M. 2006. Foreword in "Le guide des destinations indige`nes". Indige`ne editions. France: Montpellier.

Jones, S. 2005. "Community-based Ecotourism: The Significance of Social Capital." *Annals of Tourism Research*. 32: 303-324.

Kiss A. 2004. "Is community-based ecotourism a good use of biodiversity conservation funds?" *Trends in Ecology and Evolution,* 19: 232-237.

Nielsen, Erik A. 2001. Community-based Ecotourism Development and Management in the Rio Platano, *Man and the Biosphere Reserves*, Honduras.

The International Ecotourism Society (TIES). 2006. http://www.ecotourism.org/index2.php?what-is-ecotourism

Wheelen, T.L and D.J. Hunger. 2004. "Concepts in Strategic Management and Business Policy", Pearson Education, New Delhi, pp. 110, 114.

WTO, UNEP. 1992. Guidelines: Development of National Parks and Protected Areas for Tourism. *WTO/UNEP Technical Report Series* #13, Madrid, Spain.

WWF, International. 2001. Guidelines for Community-based Ecotourism Development. IUCN. (2003). Valuing Wetlands in Decision-making: Where are we now? Chapter in *Wetland Valuation Issues*. Paper 1.

10

Joint Forest Management for Sustainable Environmental Development : A Participatory Approach

Neelam Bhardwaj, Gaurav Papnai and *Y.D. Mishra*

Introduction

Since ancient times forests are the source of human sustenance and livelihood. They were the source of wildlife, timber, biodiversity, conservation of soil, water, etc. Our forefathers worshipped trees and according to an old saying "those do not have sons; trees are like sons to them". Irrespective of the ownership, forests are treated as open or community resources of local population in the sense of traditional rights. Demographic pressures after 1950s increased demands on forests and regeneration or conservation was ignored by the foresters due to consumerism. In order to reduce conflicts between local stakeholders and service providers, under National Forest Policy (1988) Government formulated a policy of involving people in the development, conservation protection and management of forest.

National Forest Policy (1988) recognized the need to involve the people in development, protection and management of forests. Creating a massive people's movement with the involvement of women became one of the basic objectives of NFP to minimize pressure of existing forests in order to ensure environmental stability and maintenance of ecological balance. The enabling circular of Ministry of Environment and Forest laid guidelines regarding the involvement of village communities and voluntary agencies in the protection, development and management of forests, giving it in the expression of Joint Forest Management. The idea was to make the people own the process and product and to empower them to change it for their common good.

There are many cases of communities protecting natural forests either on their own initiative or with the encouragement of forest department. Communities in many parts of rural India are organizing into formal and informal groups for forest protection and management in States of Orissa and Bihar, Rajasthan, Gujarat, Karnataka, Haryana, Madhya Pradesh and Punjab. While some are promoted by State Forest Departments, some are sponsored by local government or NGO programmes, e.g., village (Gram) Panchayats, women's organizations (*mahila mandals*) and tree growers cooperatives. Forest protection organizations include Groups of Village Elders, Village Forest Protection Committees, Village Councils, Village Youth Clubs, in Orissa, Forest Cooperative Societies in Kangra district of H.P., Van Panchayats in U.P. and Uttarakhand hills, Forest Protection Committees in West Bengal.

Joint Forest Management (JFM) has emerged as an important intervention in management of forest resources. Joint Forest Management emphasizes participation of local communities in an equitable partnership with the forest department. The underlying philosophy is that the local communities and the forests could coexist in a mutually sustainable manner and people could become effective keepers of the forests and their regeneration. In many parts of India, small village groups have started to protect and reclaim degraded forestlands through collective action. The Joint Forest Management Programme seeks to develop partnerships between local community institutions and State Forest Departments for sustainable management and joint benefit sharing of public forest lands. The primary objective of JFM is to ensure sustainable use of forests to meet local needs equitably while ensuring environmental sustainability. The central premise is that local women and men who are dependent on forests have the greatest stake in sustainable forest management.

Under Joint Forest Management (JFM) village communities are entrusted with the protection and management of nearby forests. These communities are required to organise forest protection committees, village forest committees, village forest conservation and development societies. The guidelines provide for rights to usufruct and non-wood forest products and percentage share of final harvest to organized communities willing to help regenerate depleted forest and wastelands.

Origin of Joint Forest Management

Joint Forest Management originated in West Bengal accidentally at the Arabari Forest Range in West Midnapore in 1971. The major hardwood of Arabari is *sal*, a commercially profitable forest crop. Ajit K. Banerjee, a silviculturist, working for the Forest Department as the Divisional Forest Officer, was conducting trials which were constantly being disturbed by grazing and illegal harvesting by the local populace. At the time there were no initiatives for sharing of forest resources between the Government and the locals, with the Government considering many of the locals no more than "thieves". The forest official, against the suggestions of his co-workers, sought out representatives of eleven local villages and negotiated the terms of a contract with an *ad hoc* Forest Protection Committee. The initial programme involved 612 families managing 12.7 square kilometres of forests classified as "degraded". 25 per cent of profits from the forests were shared with the villagers. The experiment was successful and was expanded to other parts of the State in 1987.

Sarin (1995) revealed that when villagers dependent on forests, have initiated forest protection on their own, the challenge for forest departments is of facilitating the process which implies adapting the official JFM frameworks to enable forest departments to participate in villager's initiatives. For JFM to represent participatory forest management emphasis was on regenerating timber through community protection needs to be shifted to developing sustainable alternatives for meeting diverse forest produce needs of members of community institutions. Mechanisms need to be evolved for meeting immediate essential needs of the most dependent members through appropriate forest management interventions.

Few case studies about the management and implementation of JFM projects are summarised to give insight how it is being executed in different areas of the country.

Joint Forest Management in Andhra Pradesh

Realising the importance of people's participation in forest management the "Joint Forest Management" approach was implemented in which all the fringe forest villagers were made partners in forest management with legal entitlement to the usufructs. People living on the fringes of the forests were encouraged to form *Vana Samarakshana Samithis* (VSS) to protect and rejuvenate degraded forest areas.

The *Vana Samarakshana Samithis* have been playing an important role in protection and improvement of forests. The *samithi* members with the help of micro-plans developed by them are collectively working to generate forests by removing weeds and shrubs for proper growth of important tree species in line with needs of the local people. Village development is another thrust area of VSS and focus is on improving drinking water, roads, building community halls, improving women's economic status by contributing to women's groups, etc. The *Samithis* have also been successful in upliftment of tribal and bringing them into the mainstream.

As part of support activities, smokeless *chulhas* have been provided to VSS members who have cut down consumption of firewood by more than 50 per cent. Provision of individual biogas plants has discouraged practice of storing huge stocks of firewood by villagers.

Joint Forest Management in Haryana

The JFM programme in Haryana involves communities living at the foothills of the Himalayas in the Shivaliks. Increased grazing and fuel-wood pressures accelerated the deforestation and denudation of hills. The JFM programme which started as an experiment in *Sukhomajiri* village in Shivalik hills in late 1970s, highlighted the importance of providing alternative sources of livelihood to villagers to reduce dependence on forests. The programme is now operational in 60 villages in the *Morni-Pinjore* and *Yamunanagar* forest divisions, which are organized into 57 Hill Resource Management Societies (HRMS).

The Haryana JFM programme has been successful in maintaining a balance among the three criteria, viz., of environmental soundness, affordability, and social desirability. The programme has led to a gradual shift in attitudes and actions of the Haryana Forest Department (HFD) and the HRMS towards building a partnership for the regeneration and rehabilitation of forests. Villagers under this programme have benefited from increased supply of forest produce like fodder grasses, *bhabbar* grass, irrigation water, and bamboo, which has led to increased income at household level; increased income to HRMS from sale of fodder grass, surplus *bhabbar* in open auctions, fines levied, membership fees, etc., savings; and greater employment opportunities.

Impacts include improved tree and grass cover in the forest areas, improved soil moisture regime, and reduced water run-off from the catchment. There are fewer conflicts, cases of theft, fires, illegal felling of wood, etc. With the new benefit sharing arrangements, the State derives a percentage of the profits from joint management of forest areas while a portion of the community's share is set aside for development of the forestry resources. This is to ensure the sustainability of the forest resources, and, of the JFM programme.

Joint Forest Management in Rajasthan

Rajasthan has more than 1,700 Village Forest Protection and Management Committees (VFPMC). Membership is open to one adult member willing to participate from each family of the village or village clusters. VFPMC is responsible for protection, collaborative micro-planning, production, management, resource sharing, resolution of dispute and marketing. VFPMC is required to protect and manage forest/plantation area for at least ten years to earn rights to forest produce. Before the final harvest, villagers can collect some of the non-wood forest products free of charge. While half of the income generated from auctions will go to the State exchequer 50 per cent will be deposited in the VFPMC account. Half of the share of VFPMC will be distributed equally among the members and the remaining half reinvested in plantation and forest development. In case a VFPMC does not exist, the income will go to village panchayat subject to its supportive role in protection for a minimum period of 10 years.

Participatory Forestry: Initial Promises

Case of a Rajasthan Village

Ambua, a small hamlet in Udaipur district in Rajasthan is located close to a forest block. Overuse of resources led to scarcity of fodder and livelihood goods. People felt the need to regenerate their forests and approached the forest department with their problem. Officials of the Forest Department started interacting with the villagers. A Village Forest Committee was registered. Information gathered from VFPMC meetings and PRA/RRA exercise was used for need assessment and prioritisation of needs and a detailed micro-plan for the village was prepared. Species ranking was done with the help of user groups and certain species were selected on the basis of this. The VFPMC is protecting 200 ha of plantations raised in the village and equitably distributing grass yield from the area. Villagers are entitled to take Non-Timber Forest Produce (NTFP) free of cost from the area. The committee will also get 50 per cent share in profits at the time of final harvest. The villagers collect grass free of cost from the area protected by them. VFPMC Ambua received the highest State honour—The Amrita Devi Award from the Government of Rajasthan for their work.

Villagers Protect Forests in Orissa

In Orissa 2,373 *Van Samrakshan Samiti* (VSS) are protecting 25 per cent of the total degraded forests in the State. Around 5,000 Self-Initiated Forest Protection Groups (SIFPGs) are also protecting and regenerating forests adjoining their habitation. About 78,000 ha of village woodlot plantation raised under the social forestry project is being managed by 5,600 VFCs formed by the Social Forestry Division. In *Kesharpur* village in the foothills of Orissa's *Nayagarh* district, villagers are guarding the forest from timber smugglers and stray grazing. This message has now spread to 600 other villages in the districts. Similar efforts are on in 28 out of 30 districts of Orissa.

Local communities have been protecting degraded forests, driven by economic needs. Severe forest degradation was followed by soil erosion, drying of water resources and loss of forest stripping, the villager's only source of income. There was severe drought leading to a collapse of the agriculture-based economy. The villagers realised that this was due to destruction of forests.

Sharing the minor forest produce and fuel-wood is also decided by the villagers. In *Bolangir* and *Mayurbhanj* districts villagers are free to collect *sal* leaves (for making plates), and Mahua flowers, hill broom plants and tamarind. In *Magarbhand* village, villagers pay Re. 1 for every head load of fuel-wood they collect from the protected forest and once in 3 months each household pays another rupee towards the salary of the forest guard. People are allowed into the forest for two days in a month to collect bamboo and fallen timber. Everyone is allowed to take 20 pieces of bamboo and logs for which they pay an amount fixed by the village committee.

In the *Chadayapalli* village of *Nayagarh* which protects 1,800 ha of forest land, passes are issued for timber and bamboo extraction for fuel-wood to villagers twice a week at the rate of Re. 1 per piece of bamboo. The village has been earning Rs. 90,000 every year this way, which is being used to run the school, constructing a road, etc.

In *Mayurbhanj* the money collected resulted in an institution called *Purti Society*, which now leads forest protection activities in the village. Other villages have joined and now there are 95 villages protecting the 10,000 ha of forest. The forest provides employment to villagers. Villagers are free to collect all non-timber forest produce like *sal* leaves, *mahua* and *soboi* grass (used to make ropes). Agricultural incomes of the villages have gone up. Land productivity has improved.

Conclusions from the Cases

The important conclusions drawn from the case studies are :

- JFM became the source of income for the villagers.
- The villagers could use the forest products as they were in need of these products.
- People also stopped the incidences of drought, soil erosion and other related problems.
- JFM improved tree and grass cover in the forest areas, improved soil moisture regime, and reduced water run-off from the catchment.
- Increased availability of fuel and fodder within two years of taking up JFM.
- Reduction in incidences of smuggling, fire and grazing.
- Generation of adequate employment in the village itself has reduced rural to urban migration.
- Improvement in natural regeneration of forests.
- Increase in watertable due to execution of soil and moisture conservation works.
- Increase in Non-Timber Forest Produce (NTFP) in VSS areas resulting in villagers getting more income from collection of soapnut, custard apples, etc.

According to Mishra *et al.*, (2007) conducted study in *Kumaon* region of Uttarakhand State and found that the villagers get better supply of fuelwood for preparation of food. It is easier for them to get fodder and small timbers for their animals. It also enhances the employment opportunity for the village people. These all things help in generating income of the people. Some other benefits which villagers revealed that the concept of Joint Forest Management directly or indirectly conserve the forest and sustain the environment. It provides fruits, increases aesthetic value, protect from drought and flood. Ultimately, it increases the living standard of the villagers.

Constraints in Implementation of JFM

The main aim of the JFM was to protect the forest with the participation of the local people so that they could also get the benefits from the forest. The vision behind this concept was to give local people a chance to take ownership in forest management. But it has been observed the objective with which project was visualized could not be successfully implemented. The corrupt forest officials, conflicts among the villagers due to caste and in-groups, lack of peoples' sensitivity have resulted in much hindrance for successful completion of the projects.

Mishra *et al.*, 2007 reported various constraints in implementation of JFM :

1. *Lack of Co-operation Among Villagers*: The major constraint was lack of co-operation of villagers with Village Forest Committee (VFC) members. In the village, *Pradhan* was elected as the president of VFC. People who do not like the *Pradhan* create problems and hindrance in the working of VFC. Casteism and socio-economic differences also become the matter of non-cooperation.
2. *Poor Support from Forest Department*: VFC members reported that officials of forest department create problems in releasing the funds in due time. The forest officials also do not provide cash money to the VFC members from final harvest of JFM plots which is mentioned in the agreement of the project.
3. *Illegal Grazing in JFM Area*: Villagers grazed their animals in JFM areas which caused heavy damage to the newly

planted trees. If the watchman stops them they fight with him and create problems.

4. *Fire Burning in JFM Area*: People burn the JFM areas due to jealousy with the president of VFC and other members which damages old as well as newly planted trees.
5. *Breaking of Boundary Wall*: VFC members complained that the people of the village also break the wall of the JFM area due to feeling of non-cooperation.
6. *Illegal Cutting of the Trees*: The VFC members also reported that people of nearby areas cut and sold the trees illegally.
7. *Corruption in Forest Officials*: Some of the forest department officials also indulge in the illegal cutting and sale of JFM trees. It creates an state of confusion and non-cooperation between VFC members and forest officials.
8. *Conflicts among VFC Members*: Many times the members also create conflicts due to non-cooperation and miscommunication among them. Sometimes, they blame each other for any mistake.

Strategies to Increase Participation in JFM

As Joint Forest Management is a participatory process in which contribution of each member is needed to make it sustainable. Joshi and Ghose (1999) conducted a study in which a number of suggestions were given by women for ensuring their own participation in the JFM process :

- All women forest users should become members in Gram Vikas Mandal (GVM) as it was a platform where all decisions were taken.
- Ensure Women's representation in Managing Committee of GVM, by becoming committee members.
- Make the presence of women members mandatory for decision-making.
- Discuss the forest related agenda in each MVM meeting, and convey decisions taken by women in the GVM meetings.

Here are some other strategies by which we can make more participation of the villagers in the process of forest management :

1. Awareness programme for people about forest management through participatory methods should be conducted.
2. There must be provision for all types of corruption made by anyone including VFC members, Forest officers and villagers.
3. Strengthening of Vana Samarakshana Samithis through motivation, extension, training and other support activities.
4. Forest officials must ensure equal opportunities for all villages.
5. Mobilising resources from Employment Assurance Scheme, Centrally sponsored schemes and other income-generating activities.
6. There must be regular supervision of performance of VFC by forest officials and village representatives from all sections of the village.
7. Fifty per cent compulsory enrolment for women in the general body and 30 per cent minimum in the Managing Committee should be reserved as they are the main player in the forest activities.
8. Involvement of NGOs to involve the local people in the forest management.

Conclusion

Join Forest Management is an innovative project to protect the forests and forest products. The community people are responsible for the conservation and the use of the forest products. The participation of the people and the benefits must be ensured to the local people so that this process could be strengthened. This directly protects the forests and keeps the sustainability of the environment. Joint Forest Management process has much more potential in conserving the forest reserve but for this the existing problems must be removed. JFM can become the alternative method for forest and environment development as the cases revealed around the world and the country. So there is need to promote and enhance the JFM activities in surroundings of forest areas.

REFERENCES

Arnold, J.E.M. 1992. *Community Forestry: Ten Years in Review*. Community Forestry Note of FAO, Rome.

Fisher, R.J. 1995. *Collaboration Management of Forests for Conservation and Development Issues in Forest Conservation*. IUCN, Gland.

Joshi, Seema and Bishwadeep Ghose. 1999. A participatory strategy for bringing gender equity into Joint Forest Management Exchanges, 25 and 26, June and September.

Kapoor, R.P., A.K. Upadhyaya. 1998. National Workshop on OECF assisted Forestry Projects in Jaipur during September. *Hindu*, September 17.

Mishra, Y.D., Neelam Bhardwaj and S.K. Kashyap. 2007. "Joint Forest Management (JFM) for Sustainable Development: A Study of Kumaun Region of Uttaranchal." *International Journal of Tropical Agriculture*, Vol. 25, Nos. 1-2.

Pandey, Deep Narayan. 1996. Joint Forest Management in Rajasthan: A Case Study of Three Villages in Udaipur District excerpted from *Beyond Vanishing Woods*, Himanshu/AFN, New Delhi/Udaipur.

Sarin, M. 1995. *Joint Forest Management: Achievements and Unaddressed Challenges*. Unasylva, Vol. 46 (1).

11

Sustainable Agriculture : A Part of the Solution to Current Environmental Change

Kavita Bhatnagar and *Shiwani Bhatnagar*

> *"Each one of us is a cause of global warming, but each one of us can make choices to bring our individual carbon emissions to zero. The solutions are in our hands, we just have to have the determination to make it happen."*
>
> —Al Gore

Global warming has been resulted from the increase of greenhouse gases such as carbon dioxide, methane, chlorofluorocarbons (CFC) and nitrous oxide and large-scale deforestation (Houghton, 2005). Global warming leads to global as well as local climate changes and it may potentially affect hydrospheric and biospheric environments of ecosystem and indirectly of agriculture. These global changes have some effects on agriculture such as increasing sea level due to melting of glacier, flooding of agricultural lands, unseasonal frosts, cyclones and other perturbations associated with the current climate change (Burton, 1994; Niederman, 1996). Climate change has been identified as the biggest environmental challenge that the world is facing today. Climate change has already started to alter ecosystems and agricultural conditions, and will increasingly do so over the coming decades. This will be manifested through gradual changes in rainfall amounts and patterns, temperature and the timing of seasons, all of which will affect agricultural capacity and yield. More frequent extreme weather events and the potential increase in invasive and damaging species may also significantly impact agricultural production.

It is sure that agriculture provides us with the food we all eat everyday. But do you know how those agricultural practices impact global warming? The production of agrochemicals is also an important source of greenhouse gas (GHG) emissions. In particular, fertiliser production is energy intensive, accounting for 0.6-1.2 per cent of the world's total GHGs (Bellarby *et al.*, 2008). Industrial, chemical-intensive agriculture has also degraded soils and destroyed resources that are critical to storing carbon, such as forests and other vegetation. Unsuitable agricultural activities (e.g., farm operations with heavy machines, inorganic fertilizer, animal breeding, burning of straw, soil organic carbon decline) are blamed to increase of global greenhouse gases emissions about 20 per cent (Pathak and Wassmann, 2007).

Agriculture and climate change are inextricably linked, not only will climate change affect agricultural crop production, but agriculture is a primary source of several greenhouse gases. Agricultural practices impact the global warming. Carbon dioxide is produced through several agricultural processes. The production of nitrogen fertilizer, burning of fossil fuels by machinery also emits carbon dioxide to the atmosphere. Tillage often leads to the loss of soil carbon due to enhanced decomposition of organic matter. Lime is often applied to fields to increase the soil pH, which can lead to net emissions of carbon dioxide in certain circumstances. Fertilized agricultural soils contribute a substantial amount of nitrous oxide to the atmosphere. Methane oxidation in soil is lower in agricultural soils compared to adjacent forested areas.

In order to mitigate and reduce the impacts of climate change, the first priority must be a swift and significant reduction in GHG emissions. The use of inorganic nitrogen fertilizers can be reduced by converting more land to organic farming or conventional mixed farming systems, but also by implementing judicious applications of fertilizer based on "nitrogen budgets". This would also bring great benefits for biodiversity and water quality.

According to the World Agriculture Report (IAASTD), the capacity to respond to environmental stresses can be particularly enhanced by the development and application of both traditional and local knowledge systems and new technologies. Environmental degradation constraint production and will limit the ability of agricultural systems to adapt to climatic and other global changes in many regions. Therefore, the IAASTD concludes "*sustainable agricultural practices are part of the solution to current environmental change*".

It is thus clear that agriculture needs to undergo a radical overhaul to become more sustainable. This is not just because it is important to take care of the environment, but also because sustainability is needed for the continuation of the productivity of the agro-ecosystem. Threats to the environmental sustainability of agriculture threaten agriculture itself.

The IAASTD (2008) report makes this clear by saying that greater emphasis is needed on safeguarding natural resources and agro-ecological practices, as well as on tapping the wide range of traditional knowledge held by local communities and farmers, which can work in partnership with formal science and technology. It stresses that sustainable agriculture that is biodiversity-based, including agro-ecology and organic farming, is resilient, productive, beneficial to poor farmers, and will allow adaptation to climate change.

Agriculture is thus at crossroads and it has to find ways to feed the world while being environmentally, socially and economically sustainable. Yet, it is increasingly clear that the path that agriculture has been on is not sustainable nor can it feed the world without destroying the planet (IAASTD, 2008). Once more focus is on agriculture with many critical issues that need resolving, for the agricultural development.

Sustainable agricultural approaches can be in many forms, such as agro-ecology, organic agriculture, ecological agriculture, biological agriculture, etc. Sustainable agriculture should (Pretty and Hine, 2001):

1. Make best use of nature's goods and services by integrating natural, regenerative processes, e.g., nutrient cycling, nitrogen fixation, soil regeneration and natural enemies of pests.
2. Minimise non-renewable inputs (pesticides and fertilizers) that damage the environment or harm human health.
3. Rely on the knowledge and skills of farmers, improving their self-reliance.
4. Promote and protect social capital—people's capacities to work together to solve problems.
5. Depend on locally-adapted practices to innovate in the face of uncertainty.

6. Be multifunctional and contribute to public goods, such as clean water, wildlife, carbon sequestration in soils, flood protection and landscape quality.

Sustainable agriculture, by its very definition, reduces harm to the environment, for example through the reduction or elimination of polluting substances such as pesticides and nitrogen fertilisers, water conservation practices, soil conservation practices, restoration of soil fertility, maintenance of agricultural biodiversity and biodiversity, etc.

Sustainable agricultural practices include :

- Crop rotations that alleviate weed, disease, and insect problems; increase available soil nitrogen and reduce the need for synthetic fertilisers.
- Integrated Pest Management (IPM) reduces the requisite for pesticides by crop rotations, scouting, timing of planting, biological pest controls.
- Management systems to improve plant health and crops abilities to resist pests and disease.
- Soil conserving tillage.
- Water conservation and water harvesting practices.
- Planting of leguminous crops and use of organic fertiliser or compost to improve soil fertility.

Sustainable agricultural approaches thus allow farmers to improve local food production with low-cost, readily available technologies and inputs, without causing environmental damage.

Sustainable Agriculture for Mitigating Climate Change

Importantly, sustainable agricultural practices can also mitigate climate change. Organic agriculture, for example, uses less fossil fuel based inputs and has a better carbon footprint than standard agricultural practices. This is because conventional agriculture production utilises more overall energy than organic systems due to heavy reliance on energy-intensive fertilisers and chemicals (Zeisemer, 2007). Organic agriculture performs better than conventional agriculture on a per hectare scale, both with respect to direct energy consumption (fuel and oil) and indirect consumption (synthetic fertilisers and pesticides), with high efficiency of energy use (Scialabba and Hattam, 2002).

Agriculture has the potential to change from being one of the largest GHG emitters to a net carbon sink, while offering options for mitigation. The solutions call for a shift to sustainable farming practices that build up carbon in the soil and use less fertiliser (Bellarby *et al.*, 2008). There are a variety of sustainable farming practices that can reduce agriculture's contribution to climate change, which are easy to implement. These include crop rotations and improved farming design, improved cropland management (such as avoiding leaving land bare, using an appropriate amount of fertiliser, no burning of crop residues in the field, reducing tillage), nutrient and manure management, grazing-land and livestock management, maintaining fertile soils and restoration of degraded land, improved water and rice management, and set-asides, land use change and agroforestry (Bellarby *et al.*, 2008; Niggli *et al.*, 2008).

A report by the International Trade Centre UNCTAD/WTO and FIBL (Research Institute of Organic Agriculture, Switzerland) (2007) provides a detailed assessment of the benefits of organic farming regarding climate change. The benefits are summarized as follows (Khor, 2008) :

1. Organic agriculture has considerable potential for reducing emissions.
2. In general it requires less fossil fuel per hectare and kg of produce due to the avoidance of synthetic fertilisers. Organic agriculture aims to improve soil fertility and nitrogen supply by using leguminous crops, crop residues and cover crops.
3. The enhanced soil fertility leads to stabilization of soil organic matter and in many cases to a sequestration of carbon dioxide into the soils. This in turn increases the soil's water retention capacity, thus contributing to better adaptation of organic agriculture under unpredictable climatic conditions with higher temperatures and uncertain precipitation levels. Organic production methods emphasizing soil carbon retention are most likely to withstand climatic challenges particularly in those countries most vulnerable to increased climate change. Soil erosion, an important source of carbon dioxide losses, is effectively reduced by organic agriculture.

4. Organic agriculture can contribute substantially to agroforestry production systems.
5. Organic systems are highly adaptive to climate change due to the application of traditional skills and farmers' knowledge, soil fertility-building techniques and a high degree of diversity.

Crucially, for farmers who have to face increased climate variability and extreme weather events in the near future, sustainable agriculture, by increasing resilience within the agro-ecosystem, increases its ability to continue functioning when faced with unexpected events such as climate change. For example, organic agriculture builds adaptive capacity on farms as it promotes agro-ecological resilience, biodiversity, healthy landscape management, and strong community knowledge processes (Borron, 2006). Improved soil quality and efficient water use also strengthen agro-ecosystems, while practices that enhance biodiversity allow farms to mimic natural ecological processes, which enable them to better respond to change. Sustainable farming practices that preserve soil fertility and maintain, or even increase, organic matter in soils can reduce the negative effects of drought while increasing crop productivity (Niggli *et al.*, 2008).

Consequently, there is much scope for mitigating climate change and reversing the damages through making agriculture and the food system as a whole sustainable.

REFERENCES

Bellarby, J.; Foereid, B.; Hastings, A. and Smith, P. 2008. *Cool Farming: Climate Impacts of Agriculture and Mitigation Potential.* Amsterdam: Greenpeace International.

Borron, S. 2006. *Building Resilience for an Unpredictable Future: How Organic Agriculture can help Farmers Adapt to Climate Change.* Rome: FAO.

Burton, I. 1994. "Costs of atmospheric hazards." In Proceedings of a Workshop on *Improving Responses to Atmospheric Extremes: The Role of Insurance and Compensation.* ed. D. Etkin, pp. I-1I, Theme papers. Environment Canada, Toronto.

Houghton, J. 2005. *Global Warming.* Rep. Prog. Phys. 68, 1343-1403.

IAASTD. 2008. *International Assessment of Agricultural Knowledge, Science and Technology for Development*. http://www.agassessment.org/

ITC and FiBL. 2007. *Organic Farming and Climate Change*. Geneva: International Trade Centre UNCTAD/WTO.

Khor, M. 2008. *Food Crisis, Climate Change and the Importance of Sustainable Agriculture*. Presentation at FAO Food Security Summit, Rome, 4 June, 2008.

Niederman, D. 1996. *Storm Warnings*. Report on Business Magazine, pp. 8494.

Niggli, U.; Fliessbach, A.; and Hepperly, P. 2008. *Low Greenhouse Gas Agriculture: Mitigation and Adaptation Potential of Sustainable Farming Systems*. Rome: FAO.

Pathak, H. and Wassmann, R., 2007. "Introducing Greenhouse Gas Mitigation as a Development Objective in Rice-Based Agriculture: I. Generation of Technical Coefficients." *Agricultural Systems*, 94: 807-825.

Pretty, J. and Hine, R. 2001. *Reducing Food Poverty with Sustainable Agriculture: A Summary of New Evidence*. UK: University of Essex Centre for Environment and Society.

Scialabba, N.E-H. and Hattam, C. (eds). 2002. *Organic Agriculture, Environment and Food Security*. Rome: FAO.

Zeisemer, J. 2007. *Energy Use in Organic Food Systems*. Rome: FAO.

Znaor, D.; Pretty, J.; Morrison, J.; and Todorović, S.K. 2005. *Environmental and Macroeconomic Impact Assessment of Different Development Scenarios to Organic and Low-Input Farming in Croatia*. UK: University of Essex Centre for Environment and Society.

12

Nanoscience and Nanotechnology : Social, Environmental and Human Health Issues

Chitra Singh, U.P.S. Bhardwaj and *R.P. Singh*

Introduction

Nanotechnology is a rapidly expanding field receiving extensive funding and development worldwide. Nanotechnology is a novel emerging technology that allows the manipulation of materials at the scale comparable to the size of a single molecule i.e. < 100 nanometres (nm) for its various applications to human kind. As nanoscience continues to emerge, it will impact many fields of science including sensors, biotechnology, smart materials, optics and electronics.

A.P.J. Abdul Kalam, who was the President of India during 2002-07, a distinguished technologist himself, has been stressing the role of nanotechnology as a catalyst for economic development and stability. Kalam observes: 'Nanotechnology is knocking at our doors. It is the field of the future that will replace microelectronics and many fields with tremendous application potential in the areas of medicine, electronics, and material science' (Kalam, 2006). Nanoscience and nanotechnologies are widely seen as having huge potential to bring benefits in areas as diverse as drug development, water decontamination, information and communication technologies, and production of stronger and lighter materials. A large number of novel processes occur on the nanoscale, and the unique properties of nanoparticles offer potential benefits in a range of applications from consumer products to medicine and environmental protection. The large surface area to volume ratio of nanofibre membranes allows greater surface adsorption of contaminants for air and water and increases the lifetime of filtration media. However, there remains considerable uncertainty with regard to the hazards and risks associated with nanoscience.

Environmental Implication of Nanotechnology

Exposure of nanomaterials to workers, consumers and the environment seems inevitable with the increasing production volumes and the increasing number of commercially available products containing nanomaterials based on nanotechnology. There are potential toxicological and eco-toxicological effects of nanoparticles through exposure. Exposure of nanoparticles in humans may take place through various routes such as dermal, inhalation, ingestion and injection. Human exposure of nanoparticles can take up through infected/infested living organisms in water/sediment or through food chain and ingestion of nanoparticles sorbed on vegetation, etc.

Free nanoparticles could get into environment through various routes such as its discharge in waste water from research laboratories; companies manufacturing nanoparticles based drugs, cosmetics, creams, silver, titanium oxide, etc. Another route of environmental exposure of nanoparticles are spills from production, transport and disposal are waste water overflow nanomaterials or products.

Studies of certain nanoparticles on specific test species on brain, gills, liver, intestine and also with embryonic development of eggs. It was observed that various combinations and concentration of (Fullerenes, C 60) caused lipid peroxidation in brain, gills and levers which was attributed to reparation enzymes (Oberdorster, 2004; Smith *et al.*, 2007) which may ultimately even lethal. (Smith *et al.*, 2007) observed a dose dependent rise in ventilation rate, mucous secretion on the gill and total glutathione levels in the gills and liver.

An enhancement of inherent toxicity, chemical or biological reactivity have been observed due to increased surface area in relation to mass for smaller nanoparticles compared to larger molecule (Oberdorster, 2004). Inhalation studies on rodents have found that ultra fine particles of titanium dioxide causes larger lung damage in rodents compared to larger fine particles for the same amount of the substances. Nanoparticles may consist of transition metals, metal oxides, carbon based nanomaterials. The structure of nanoparticles has been shown to have a profound influence on the toxicity. The study also reveals that single walled carbon nanotubes were more toxic compared to multi-walled C nanotubes as well as C-60 (Oberdorster *et al.*, 2005).

Nanopollution

The environmental impacts of nanotechnology have become increasingly area of research. There are also dangerous nanosized particles and chemicals such as asbestos, diesel particulate matter, ultra fine particles, DDT, lead, etc. which can accumulate in the food chain. Nanopollution is a generic name for all waste generated by nanodevices or during the process of manufacturing of nanomaterials. It can float in the air and might easily penetrate into plant and animal cells because of its tiny size causing unknown effects.

Environmental Assessment of Nanotechnology

Eco-toxicological impact of nanoparticles and the potential for its bioaccumulation and transport mechanism has to be thoroughly studied. In free form nanoparticles can be released in the air or water during production or as a by-product of production and ultimately accumulate in the soil, water or plant life. Manufactured nanoparticles will ultimately have to be recycled or disposed of as waste. However, it is yet not known whether these nanoparticles are a new class of non-biodegradable pollutant.

Environmental Application of Nanotechnology

Nanotechnology plays a major role in the development of new products to substitute existing production equipments, materials and chemicals with improved performance resulting in potential environmental cost savings. Reduced consumption of materials also benefits the environment.

Energy

Nanotechnology is a fast emerging key enabling technology that is often cited as the solution to achieving material and energy efficiency

(Masciangioli and Zhang, 2003; Colvin, 2003). Nanotechnology could potentially have a great impact on clean energy production. Research is underway to use nanomaterials in solar cells, fuel cells and environmental friendly batteries. It is believed that these nanomaterials based electronics devices will enable more efficient solar cells and would have a great effect on satisfying global energy needs. Another example of environmental friendly form of energy is the use of fuel cells powered by hydrogen. The use of nanomaterials may enable batteries with higher energy content or super capacitors with higher rate of recharging which could be helpful for frequent battery disposal problem.

Water Filtration and Remediation

Nanoporous membranes which may be composed of nanotubes are suitable for mechanical filtration with extremely small pores less than 10 nm. Nano filtration is mainly used for the removal of ions or the separation of different fluids. The nanofibres used in filtration membranes for environmental remediation, minimizing the pressure drop and provide better efficiency than conventional fibre mats. The large surface area-to-volume ratio of nanofibre membranes allows greater surface adsorption of contaminants for air and water and increases the lifetime of filtration media. Magnetic nanoparticles offer an effective and reliable method to remove heavy metal contaminants from waste water by making use of magnetic separation techniques. Nanoscale particles increases the efficiency to absorb the contaminants and is comparatively inexpensive compared to traditional precipitation and filtration methods.

Social Issues

Nanotechnology has the potential to change and improve many sectors of our society, from consumer products to health care, transportation, energy and agriculture. It also affects the global economy through research and development of nanomaterials into production and marketing. However, the introduction of large quantities of nanostructured materials into our society may have unforeseen consequences for environmental quality. In order to serve the needs of the society, the technology has to be used in a favourable socio-economic context. For the poor, socio-economic structure is a difficult barrier than technological innovation.

Nanotechnology is able to revolutionise manufacturing, health care, transportation, energy and food supply, medical system, the military, the agricultural enterprise and also even to transform labour and workplace. Each one of these technology dependant sector is operated by and for society. Nanotechnology has potential to produce easy-to-handle devices which reduce the dependence on human skills on many fronts, thus reducing the human risk.

Conclusion

Nanotechnology can assist resource saving through the use of light weight, high strength materials based on carbon nanotubes and metal oxide frameworks as hydrogen storage material. Nanotechnology is the manipulation our self-assembly of individual atoms, molecules, or molecular clusters into structures to create materials and devices with new or vastly different properties. Nanotechnology has the potential to revolutionize the agricultural and food industry with new tools for the molecular treatment of diseases, rapid disease detection, enhancing the ability of plants to absorb nutrients, etc. Nanotechnology will also protect the environment indirectly through the use of alternative (renewable) energy supplies, and filters or catalysts to reduce pollution and clean up existing pollutants. Nanotechnology can also be used to clean groundwater for drinking purpose. Nanotechnology can do a lot to benefit society through its application in agriculture industry, Environmental protection, and medical science and food systems.

REFERENCES

Colvin, V.L. 2003. "The potential environmental impact of engineered nanomaterials." *Nat. Biotech.* 21(10): 1166-1170.

http://stm.dev.pennet.com/articles/article_display.cfm?Section=ARCHI&C=Finan&ARTICLE_ID=269635&p=109

Kalam, A.P.J. 2006. Quoted in Stuart, C. "India sees growth through nanotech." *Small Times*.

Masciangioli, T. and W. Zhang. 2003. "Environmental technologies at the nanoscale." *Env. Sci. Tech.* 37950: 102A-108A.

Oberdorster, E. 2004. "Manufactured nanomaterials (Fullerenes, C-60) induce oxidative stress in the brain of juvenile largemouth bass." *Environmental Health Perspectives*, 112 (10), 1058-1062.

Oberdorster, G.; E. Oberdorster; J. Oberdorster. 2005. "Nanotoxicology: An emerging discipline evolving from studies of ultra fine particles." *Environmental Health Perspective*, 113 (7), pp. 823-839.

Smith, C.J.; B.J. Shaw; R.D. Handy. 2007. "Toxicity of Single Walled Carbon Nanotubes on Rainbow Trout (Oncorhynchus mykiss): Respiratory Toxicity, Organ Pathologies, and Other Psychological Effects." *Aquatic Toxicology*.

13

Green Marketing : A Conceptual Approach for Solving Environmental Problems

Nisha Singh, Deepika Singh, Pratibha Singh, Ashima Srivastava and *Roli Verma*

Introduction

The twentieth century was a time of unparalleled growth. The world's population grew to pass the six billion mark. We witnessed the birth of the mass market and the boom in mass production to satisfy its needs. As markets became more competitive, so formal marketing emerged and became increasingly important as a means by which companies could continue to grow their markets and their market shares. The comfortable assumption was that the all out pursuit of economic growth was the most beneficial strategy for development, because the wealth generated could be invested to improve the quality of life of those inside and outside the industrialized economies. At the beginning of the twenty-first century, the social and environmental consequences of the unquestioning pursuit of economic growth have become increasingly clear. Increasing levels of greenhouse gases in the atmosphere, a hole in the ozone layer caused by CFC releases, widespread destruction of the rain forests and a growing list of endangered species and ecosystems are just a few of the indicators that all is not well. By 2000, World Bank figures showed that nearly half the world's population lives on under \$2 per day. For this half of the world, issues of consumer choice and sovereignty or discretionary spending have little meaning, and promises that the growth in the industrialized economies would lead to a better quality of life for them have generally not been fulfilled. For the new century, the key challenge for mankind is to find more sustainable and equitable ways to produce, consume and live. Sustainability was once a vision of the future shared by an environmentally-orientated few. The publication of the Brundtland Report 'Our Common Future' in 1987 brought the issue into the mainstream. In the wake of the 1992 Rio Earth Summit, the world's governments and major corporations have increasingly adopted the pursuit of sustainability as a goal. For marketing, the challenge is two-fold. In the short term, ecological and social issues have become significant external influences on companies and the markets within which they operate. Companies have to react to changing customer needs, new regulations and a new social Zeitgeist which reflects increasing concern about the socio-environmental impacts of business. In the longer term, the pursuit of sustainability will demand fundamental changes to the management paradigm which underpins marketing and the other business functions. This article aims to illustrate how the 'green challenge' is exerting an influence on current marketing practice and how its implications will require a more profound shift in the marketing paradigm, if marketers are to continue delivering customer satisfaction at a profit throughout this new millennium (Crane, 2000).

Green Marketing in Context

Management theory in general is firmly rooted in an economic and technical systems perspective which concentrates on exchanges, products, production and profits. Over time it has evolved to become more 'human', with the emergence of disciplines like organizational behaviour, human resource management, business ethics and societal marketing. The fact that businesses are physical systems which exist within a finite and vulnerable physical environment has, until recently, largely been ignored as a management and marketing issue. During the 1990s, the marketing discipline began to seriously discuss the physical implications and sustainability of marketing. The roots of green marketing can be traced back to the wave of environmental concern of the 1970s which spawned the 'ecological marketing' concept. This was largely concerned with those industries with the most severe environmental impacts, and with developing new technologies to alleviate particular environmental problems. The reaction of many marketing academics and practitioners to the environmental concern which emerged in the late 1980s and early 1990s was, therefore, a feeling of *déjà vu*. The integration of environmental concern into marketing theory and practice is also viewed by some as simply an extension of the societal marketing concept. However, it is more helpful to view the concept of 'green marketing' which has developed as a response to the green movement of today, as something which integrates and expands upon the ideas embedded in the ecological and societal marketing concepts. It can be defined as: The holistic management process responsible for identifying, anticipating and satisfying the needs of customers and society, in a profitable and sustainable way. The key differences between the green marketing concept and societal marketing lie in:

- An emphasis on the physical sustainability of the marketing process, as well as its social acceptability.
- A more holistic and interdependent view of the relationship between the economy, society and the environment.
- An open-ended rather than a long-term perspective.
- A treatment of the environment as something with intrinsic value over and above its usefulness to society.
- A focus on global concerns, rather than those of particular societies.

In both the ecological and societal marketing concepts, the emphasis on socio-environmental issues for marketeers has mostly been framed in terms of costs and constraints. Another important new dimension that green marketing introduced was an emphasis on socio-environmental issues as a potential source of innovation and opportunity for marketeers. Combining environmental concern (which traditionally involves encouraging conservation), with the discipline of marketing (which aims to stimulate and facilitate consumption) can appear somewhat paradoxical. Sustainability is the keystone of the green marketing philosophy, which resolves this apparent paradox. A sustainable approach to consumption and production involves enjoying a material standard of living today, which is not at the expense of the standard of living of future generations. It is a deceptively simple concept comprising two parts:

- Using natural resources at a rate at which environmental systems or human activity can replenish them (or in the case of non-renewable resources, at a rate at which renewable alternatives can be substituted in).
- Producing pollution and waste at a rate which can be absorbed by environmental systems without impairing their viability.

Green marketing's key concepts of sustainability and holism are both apparently simple, but can be extremely difficult to translate into action. This is largely because conventional management wisdom emphasizes reductionalism and specialism, and is founded on economic theories which mistakenly treat environmental resources as limitless, free (beyond the cost of extraction) or, for market-less commodities like stratospheric ozone, worthless (Elkington, 1994).

An important role in marketing green products is to minimize the sacrifices that consumers must make in order to buy and use 'green' products. Improvements must be substantive, and if possible backed by hard facts.

Different environmental attributes are important in different product categories. Nevertheless, consumers generally feel that the more attributes there are, the better. A good strategy is to offer environmental improvements in several categories at the same time, such as water pollution, solid waste, and manufacturing processes.

When consumers are faced with a trade-off between product attributes, the environment almost always loses. Many products that require consumers to make such trade-offs have failed to establish themselves in the marketplace.

To succeed, eco-entrepreneurs must not neglect the traditional values of price, quality, convenience and availability.

- *Price*: No matter what consumers say, few will pay extra for greener products. Premiums of more than around 2 per cent can be charged only if the consumer perceives additional product value.
- *Quality*: Many people think 'green' products will be less effective than existing brands, and are even less likely to sacrifice quality than their price. Assurances of quality are absolutely essential, and must be communicated

convincingly. Quality is judged on many features, including performance, look, feel, fit, comfort, and durability.

- *Convenience*: Only minimal inconvenience is tolerated by consumers of green products.
- *Availability*: Very few consumers will go out of their way to buy green products. In a 1990 poll, 58 per cent of respondents cited poor product availability as a reason for not purchasing more green products.

So-called 'deep-green' markets are too small to attract major competition, which may be an important factor in planning long-term strategy. Recent estimates have put the size of this market at about 2 per cent of North American consumers. These individuals will often pay higher prices or sacrifice product performance for environmental improvements.

If you decide to target this market,

- Your products must be significantly better, in environmental terms, than the competition, or else you will be viewed as a hypocrite as well as losing the sale;
- You must provide clear, persuasive and comprehensive information on what sets your product apart environmentally;
- You should monitor competitors to see whether they are developing products almost as green as yours, but without the price and quality trade-offs. You can keep ahead of the green development curve by minimizing the sacrifices required to purchase or use your product.

Both traditional and 'green' companies are beginning to offer consumers environmental improvements that do not involve a price premium or a loss of quality.

Sharp-eyed marketing people in the late 1980s discovered a new advertising ploy: the green claim. In 1993, some 13 per cent of new products introduced in the US made some kind of environmental claim. But some claims were misleading and served to confuse, rather than assist, consumers. Canadian consumers report that their distrust of green product claims inhibited their purchases of allegedly 'green' products.

In July 1992, the US Federal Trade Commission issued guidelines on the use of such terms as 'degradable', 'compostable', 'recyclable', 'recycled content', 'refillable' and 'ozone safe'.

The message is clear: In order to be successful, products and services must be credible to cynical and confused consumers. One way to gain credibility is to obtain certification from one of the following organizations :

- Scientific Certification Systems of Oakland, California was founded in 1984 to measure pesticide residues in agricultural products. It now certifies a wide array of environmental and food safety claims. Under its Environmental Claims Certification Programme, introduced in 1990, SCS has certified single-criteria claims such as 'recycled content', 'biodegradable', 'water-efficient', and 'no VOCs' for more than 1,100 products.

The SCS 'Environmental Report Card', launched in 1993, provides detailed information about the performance and environmental impacts of a product, based on a life-cycle, cradle-to-grave study of the product and its packaging. The impacts listed on the card include resources depleted, energy consumed, pollution released into air and water, and solid waste generated.

Among the products for which report cards have been commissioned, at a cost of up to $30,000 each, are American Plastics 'IronHold' trash bags; 'PlastiKote' enamel paints; Benckiser 'EarthRite' household cleaning products; and 'TreeFree' recycled napkins, paper towels, and tissues (Mintel, 1991).

Many people believe that green marketing refers solely to the promotion or advertising of products with environmental characteristics. Generally terms like Phosphate Free, Recyclable, Refillable, Ozone Friendly, and Environmentally Friendly are some of the things consumers most often associate with green marketing. In general green marketing is a much broader concept, one that can be applied to consumer goods, industrial goods and even services. For example, around the world there are resorts that are beginning to promote themselves as "ecotourism" facilities, i.e., facilities that specialize in experiencing nature or operating in a fashion that minimizes their environmental impact. Thus green marketing incorporates a broad range of activities, including product modification, changes to the production process, packaging changes, as well as modifying advertising. But to define green marketing is not a simple task. The terminology used in this area has varied, it includes: Green Marketing, Environmental Marketing and Ecological Marketing. While green marketing came into prominence in the late 1980s and early 1990s, it was first discussed much earlier. The American Marketing Association (AMA) held the first workshop on "Ecological Marketing" in 1975. The proceedings of this workshop resulted in one of the first books on green marketing entitled *Ecological Marketing*.

Definition of Green Marketing

Green marketing is defined as *"Green or Environmental Marketing consists of all activities designed to generate and facilitate any exchanges intended to satisfy human needs or wants, such that the satisfaction of these needs and wants occurs, with minimal detrimental impact on the natural environment."* Green marketing symbol is shown in Figure 13.1.

This definition incorporates much of the traditional components of the marketing definition, that is "All activities designed to generate and facilitate any exchanges intended to satisfy human needs or wants." Therefore, it ensures that the interests of the organization and all its consumers are protected, as voluntary exchange will not take place unless both the buyer and seller mutually benefit. The above definition also includes the protection of the natural environment, by attempting to minimize the detrimental impact this

Fig. 13.1 : Green Marketing Symbol

exchange has on the environment. This second point is important, for human consumption by its very nature is destructive to the natural environment. So green marketing should look at minimizing environmental harm, not necessarily eliminating it.

Advantages

Though human needs are unlimited, resources are limited. It is significant for the marketers to employ the resources in an efficient way without misuse as well as to attain the objective of the organization. It is proved that green marketing is unavoidable. The customers will easily be able to see the 'greenness' of that particular product or service. It is growing rapidly. The interest of the consumers is also growing all-over the world to save the environment.

- *Product* : The products should be made in such a way that it can help the customers who like environment-friendly products. It can be made from recycled materials or from used things.

Fig 13.2: Blue Globe with Green Check

- *Price* : You have to be careful about green pricing.
- *Place* : Green place is all about handling logistics to bring down on transportation emissions.
- *Promotion* : It includes several things such as advertising, marketing objects, signage, white papers, websites, videos and presentations.

Elements Required for Green Marketing

Three basic elements are required for a thriving green marketing campaign:

- *Credibility* : It is the key element in the field of marketing. You have to acquire credibility from your target audience. You do not have to do away with all the environment-unfriendly practices; you should unveil them sooner or later. While trying to create your credibility, you should look for a convincing third-party certification. It can come from private or non-governmental organizations.
- *Address a Requirement* : This is the key objective of any business. Your product or provided services should tell about its necessity. Green products and services do not obtain a pass as they have environment-friendly features. Most of the customers are reluctant to pay more for these types of products or services. If you are going to include any green feature into an existing product, you have to be confident of your claims. If you are making a new green products or services, first make out the need and then address it in an environment-friendly way.
- *Converse using Eco-Friendly Channels*: You have to be inventive while you are going to promote your product or service. Look into new channels like solar power billboards or street advertising.

Green marketing covers more than the marketing claims of an organization (Ottman, 1993). The SCS Forest Conservation Programme, launched in 1991, uses an indexing system to rank forest management practices. To raise consumers' awareness of the environmental impact of their own daily activities, 'Personal Environmental Report Card' has been developed. The evaluation includes commuting distances and transportation methods, home energy and water use, and recycling habits.

- Green Seal is a US based, non-profit environmental labelling organization which awards a seal of approval to products that are found to cause significantly less harm to the environment than other, similar products. Green Seal develops environmental standards for consumer products through a public review process involving manufacturers, environmentalists, consumers and Government agencies.

Products are Green Seal certified only after rigorous testing and evaluation. The primary testing contractor is Underwriters Laboratories. The Green Seal Certification Mark—a blue globe with a green check—assures consumers that products meet stringent environmental standards.

Green Seal standards take into consideration each product's environmental impact, measured over the whole life cycle. By setting standards for environmentally preferable products, Green Seal is designed to reduce air and water pollution, reduce the wastage of energy and natural resources, curb greenhouse gas emissions, prevent toxic contamination, and protect wildlife.

Green Seal has developed standards for more than 50 product categories, including paints, household cleaners, paper products, water-efficient fixtures, lighting, windows, and major household appliances.

- Canada's government-sponsored 'Environmental Choice' programme started by evaluating single-criteria product claims, but has subsequently adopted a life-cycle review process. Certified items carry the EcoLogo—three doves intertwined to form the shape of a maple leaf.

The logo helps consumers identify products that, in the course of their life cycle, improve energy efficiency, use recycled or recyclable materials, and minimize the use of hazardous substances. A 16-member national board provides advice on the development of environmental criteria, and an independent technical agency assists in the verification and testing of products and services. Lists of certified brand name products and their manufacturers are made available for consumers.

Product categories evaluated to date include re-refined motor oil, recycled fine paper, water-based paints, recycled newsprint, thermal insulation, ethanol-blended gasoline, reusable diapers, alkaline batteries, composters for residential waste, water-saving products, laundry detergents, domestic water heaters, toner cartridges, energy-efficient lamps, and adhesives.

The obvious assumption of green marketing is that potential consumers will view a product or service's "greenness" as a benefit and base their buying decision accordingly. The not-so-obvious assumption of green marketing is that consumers will be willing to pay more for green products than they would for a less-green comparable alternative product—an assumption that, in my opinion, has not been proven conclusively (Porter and Van der Linde, 1995).

While green marketing is growing greatly as increasing numbers of consumers are willing to back their environmental consciousnesses with their dollars, it can be dangerous. The public tends to be skeptical of green claims to begin with and companies can seriously damage their brands and their sales if a green claim is discovered to be false or contradicted by a company's other products or practices. Presenting a product or service as green when it is not called greenwashing. Examples: Chad's green marketing campaign bombed because he made the mistake of packaging his environmental-friendly product in styrofoam.

Green Marketing related Terminology

A. Green Product

The products have to be developed depending on the needs of the customers who prefer environment-friendly products. Products can be made from recycled materials or from used goods. Efficient products not only save water, energy and money, but also reduce harmful effects on the environment. Green chemistry forms the growing focus of product development. The marketer's role in product management includes providing product designers with market-driven trends and customer requests for green product attributes such as energy saving, organic, green chemicals, local sourcing, etc. For example, Nike is the first among the shoe companies to market itself as green. It is marketing its Air Jordan shoes as environment-friendly, as it has significantly reduced the usage of harmful glue adhesives. It has designed this variety of shoes to emphasize that it has reduced wastage and used environment-friendly materials.

B. Green Price

Green pricing takes into consideration the people, planet and profit in a way that takes care of the health of employees and communities and ensures efficient productivity. Value can be added to it by changing its appearance, functionality and through customization, etc. Wal Mart unveiled its first recyclable cloth shopping bag. IKEA started charging consumers when they opted for plastic bags and encouraged people to shop using its "Big Blue Bag".

C. Green Place

Green place is about managing logistics to cut down on transportation emissions, thereby in effect aiming at reducing the carbon footprint. For example, instead of marketing an imported mango juice in India it can be licensed for local production. This avoids shipping of the product from far away, thus reducing shipping cost and more importantly, the consequent carbon emission by the ships and other modes of transport.

D. Green Promotion

Green promotion involves configuring the tools of promotion, such as advertising, marketing materials, signage, white papers, websites, videos and presentations by keeping people, planet and profits in mind. British Petroleum (BP) displays gas station which its sunflower motif and boasts of putting money into solar power. Indian Tobacco Company has introduced environment-friendly papers and boards, which are free of elemental chlorine. Toyota is trying to push gas/electric hybrid technology into much of its product line. It is also making the single largest R&D investment in the every-elusive hydrogen car and promoting itself as the first eco-friendly car company. International Business Machines Corporation (IBM) has revealed a portfolio of green retail store technologies and services to help retailers improve energy efficiency in their IT operations. The centre piece of this portfolio is the IBM SurePOS 700, a point-of-sale system that, according to IBM, reduces power consumption by

36 per cent or more. We even see the names of retail outlets like "Reliance Fresh", Fresh@Namdhari Fresh and Desi, which while selling fresh vegetables and fruits, transmit an innate communication of green marketing.

Green marketer can attract customers on the basis of performance, money savings, health and convenience, or just plain environmental friendliness, so as to target a wide range of green consumers.

Consumer awareness can be created by spreading the message among consumers about the benefits of environment-friendly products. Positing of profiles related to green marketing on social networks creates awareness within and across online peer groups. Marketing can also directly target the consumers through advertisements for product such as energy saving compact fluorescent lamps, the battery-powered Reva car, etc. (Shelton, 1994).

Importance of Green Marketing

Man has limited resources on the earth, with which she/he must attempt to provide for the worlds' unlimited wants. There is extensive debate as to whether the earth is a resource at man's disposal. In market societies where there is "freedom of choice", it has generally been accepted that individuals and organizations have the right to attempt to have their wants satisfied. As firms face limited natural resources, they must develop new or alternative ways of satisfying these unlimited wants. Ultimately green marketing looks at how marketing activities utilize these limited resources, while satisfying consumers' wants, both of individuals and industry, as well as achieving the selling organization's objectives.

When looking through the litrature there are several suggested reasons for firms' increased use of Green Marketing. Five possible reasons are as follows :

1. Organizations perceive environmental marketing to be an opportunity that can be used to achieve its objectives.
2. Organizations believe they have a moral obligation to be more socially responsible. Governmental bodies are forcing firms to become more responsible.
3. Competitors' environmental activities pressure firms to change their environmental marketing activities.

4. Cost factors associated with waste disposal, or reductions in material usage forces firms to modify their behaviour. (Smithe, 1998).

Opportunities

All types of consumers, both individual and industrial are becoming more concerned and aware about the natural environment. In a 1992 study of 16 countries, more than 50 per cent of consumers in each country, other than Singapore, indicated they were concerned about the environment. A 1994 study in Australia found that 84.6 per cent of the sample believed all individuals had a responsibility to care for the environment. A further 80 per cent of this sample indicated that they had modified their behaviour, including their purchasing behaviour, due to environmental reasons. As demands change, many firms see these changes as an opportunity to be exploited. It can be assumed that firms marketing goods with environmental characteristics will have a competitive advantage over firms marketing non-environmentally responsible alternatives. There are numerous examples of firms who have strived to become more environmentally responsible, in an attempt to better satisfy their consumer needs. McDonald's replaced its clam shell packaging with waxed paper because of increased consumer concern relating to polystyrene production and ozone depletion. Xerox introduced a "high quality" recycled photocopier paper in an attempt to satisfy the demands of firms for less environmentally harmful products.

This is not to imply that all firms who have undertaken environmental marketing activities actually improve their behaviour. In some cases firms have misled consumers in an attempt to gain market share. In other cases firms have jumped on the green bandwagon without considering the accuracy of their behaviour, their claims, or the effectiveness of their products. This lack of consideration of the true "greenness" of activities may result in firms making false or misleading green marketing claims.

Social Responsibility and Green Marketing

Many firms are beginning to realize that they are members of the wider community and, therefore, must behave in an environmentally responsible fashion. This translates into firms that believe they must achieve environmental objectives as well as profit related objectives. This results in environmental issues being integrated into the firm's corporate culture. Firms in this situation can take two perspectives: (1) they can use the fact that they are environmentally responsible as a marketing tool; or (2) they can become responsible without promoting this fact. There are examples of firms adopting both strategies. Organizations like the Body Shop heavily promote the fact that they are environmentally responsible. While this behaviour is a competitive advantage, the firm was established specifically to offer consumers environmentally responsible alternatives to conventional cosmetic products. This philosophy is directly tied to the overall corporate culture, rather than simply being a competitive tool. An example of a firm that does not promote its environmental initiatives is Coca-Cola. They have invested large sums of money in various recycling activities, as well as having modified their packaging to minimize its environmental impact. While being concerned about the environment, Coke has not used this concern as a marketing tool. Thus many consumers may not realize that Coke is a very environmentally committed organization. Another firm who is very environmentally responsible but does not promote this fact, at least outside the organization, is Walt Disney World (WDW). WDW has an extensive waste management programme and infrastructure in place, yet these facilities are not highlighted in their general tourist promotional activities.

Governmental Pressure

As with all marketing related activities, governments want to "protect" consumers and society; this protection has significant green marketing implications. Governmental regulations relating to environmental marketing are designed to protect consumers in several ways: (1) reduce production of harmful goods or by-products; (2) modify consumer and industry's use and/or consumption of harmful goods; or (3) ensure that all types of consumers have the ability to evaluate the environmental composition of goods.

Governments establish regulations designed to control the amount of hazardous wastes produced by firms. Many by-products of production are controlled through the issuing of various environmental licenses, thus modifying organizational behaviour. In some cases governments try to "induce" final consumers to become more responsible. For example, some governments have introduced voluntary curb-side recycling programmes, making it easier for consumers to act responsibly. In other cases governments tax individuals who act in an irresponsible fashion. For example in Australia there is a higher gas tax associated with leaded petrol.

Competitive Pressure

Another major force in the environmental marketing area has been firms' desire to maintain their competitive position. In many cases firms observe competitors promoting their environmental behaviours and attempt to emulate this behaviour. In some instances this competitive pressure has caused an entire industry to modify and thus reduce its detrimental environmental behaviour. For example, it could be argued that Xerox's "Revive 100 per cent Recycled paper" was introduced a few years ago in an attempt to address the introduction of recycled photocopier paper by other manufacturers. In another example when one tuna manufacturer stopped using driftnets the others followed suit.

Cost or Profit Issues

Firms may also use green marketing in an attempt to address cost or profit related issues. Disposing of environmentally harmful by-products, such as polychlorinated biphenyl (PCB) contaminated oil are becoming increasingly costly and in some cases difficult. Therefore, firms that can reduce harmful wastes may incur substantial cost savings. When attempting to minimize waste, firms are often forced to re-examine their production processes. In these cases, they often develop more effective production processes that not only reduce waste, but reduce the need for some raw materials. This serves as a double cost savings, since both waste and raw material are reduced. In other cases firms attempt to find end-of-pipe solutions, instead of minimizing waste. In these situations firms try to find markets or uses for their waste materials, where one firm's waste becomes another firm's input of production. One Australian example of this is a firm who produces acidic waste water as a by-product of production and sells it to a firm involved in neutralizing base materials. The last way in which cost or profit issues may affect firms' environmental marketing activities is that new industries may be developed. This can occur in two ways: (1) a firm develops a technology for reducing waste and sells it to other firms; or (2) a waste recycling or removal industry develops. For example, firms that clean the oil in large industrial condensers increase the life of those condensers, removing the need for replacing the oil, as well as the need to dispose of the waste oil. This reduces operating costs for those owning the condensers and generates revenue for those firms cleaning the oil.

Some Problems with Green Marketing

There are a number of potential problems that must overcome. One of the main problems is that firms using green marketing must ensure that their activities are not misleading to consumers or industry, and do not breach any of the regulations or laws dealing with environmental marketing. Green marketing claims must clearly state environmental benefits. A problem of the firms face is that those who modify their products due to increased consumer concern must contend with the fact that consumers' perceptions are sometimes not correct. For example the McDonald's case where it has replaced its clam shells with plastic coated paper. There is ongoing scientific debate which is more environmental-friendly. Some scientific evidence suggests that when taking a cradle to grave approach, polystyrene is less environmentally harmful if this is the case McDonald's bowed to consumer pressure, yet has chosen the more environmentally harmful option. When firms attempt to become socially responsible, they may face the risk that the environmentally responsible action of today will be found to be harmful in the future. Take for example the aerosol industry which has switched from CFCs (chlorofluorocarbons) to HFCs (hydro fluorocarbons) only to be told HFCs are also a greenhouse gas. Some firms now use DME (di-methyl ether) as an aerosol propellant, which may also harm the ozone layer. Given the limited scientific knowledge at any point, it may be impossible for a firm to have made the correct environmental decision. This may explain why some firms, like Coca-Cola and Walt Disney World, are becoming socially responsible without publicizing the point. They may be protecting themselves from potential future negative backlash; if it is determined they made the wrong decision in the past. While governmental regulation is designed to give consumers the opportunity to make better decisions or to motivate them to be more environmentally responsible, there is difficulty in establishing policies that will address all environmental issues. For example, guidelines developed to control environmental marketing address only a very narrow set of issues, i.e., the truthfulness of environmental marketing claims. If governments want to modify consumer behaviour they need to establish a different set of regulations. Thus governmental attempts to protect the environment may result in a proliferation of regulations and guidelines, with no one central controlling body. Reacting to competitive pressures can cause all "followers" to make the same mistake as the "leader." Mobil Corporation who has followed the competition and introduced "biodegradable" plastic garbage bags, as because technically these bags were biodegradable, the conditions under which they were disposed did not allow biodegradation to occur. Mobil was sued by several US states for using misleading advertising claims. Thus blindly following the competition can have costly ramifications.

The push to reduce costs or increase profits may not force firms to address the important issue of environmental degradation. End-of-pipe solutions may not actually reduce the waste but rather shift it around. While this may be beneficial, it does not necessarily address the larger environmental problem, though it may minimize its short term affects. Ultimately, most waste produced will enter the waste stream, therefore, to be environmentally responsible organizations should attempt to minimize their waste, rather than find "appropriate" uses for it. (Wong *et al.*, 1996)

Conclusion

Green marketing covers more than a firm's marketing claims. While firms must bear much of the responsibility for environmental degradation, ultimately it is consumers who demand goods, and thus create environmental problems. One example of this is where McDonald's is often blamed for polluting the environment because much of their packaging finishes up as roadside waste. It must be remembered that it is the uncaring consumer who chooses to dispose of their waste in an inappropriate fashion. While firms can have a great impact on the natural environment, the responsibility should not be theirs alone. It appears that consumers are not overly committed to improving their environment and may be looking to lay too much responsibility on industry and Government. Ultimately, green marketing requires that consumers want a cleaner environment and are willing to "pay" for it, possibly through higher priced goods, modified individual lifestyles, or even governmental intervention. Until this occurs it will be difficult for firms alone to lead the green marketing revolution. It must not be forgotten that the industrial buyer also has the ability to pressure suppliers to modify their activities. Thus, an environmental committed organization may not only produce goods that have reduced their detrimental impact on the environment, they may also be able to pressure their suppliers to behave in a more environmentally "responsible" fashion. Final consumers and industrial buyers also have the ability to pressure organizations to integrate the environment into their corporate culture and thus ensure all organizations minimize the detrimental environmental impact of their activities.

REFERENCES

Crane, A. 2000. "Facing the backlash: Green marketing and strategic re-orientation in the 1990s." *Journal of Strategic Marketing*, Vol. 8, No. 3, pp. 277-96.

Elkington, J. 1994. "Towards the sustainable corporation: Win business strategies for sustainable development." *California Management Review*, Vol. 36, No. 2, pp. 90-100.

Mintel. 1991. The Green Consumer Report, London.

Ottman, J.A. 1993. *Green Marketing: Challenges and Opportunities*, NTC Business Books, Chicago, IL.

Porter, M.E.; Linde Van der. 1995. "Green and competitive: Ending the stalemate." *Harvard Business Review,* Vol. 73, No. 5, pp. 120-33.

Shelton, R.D. 1994. "Hitting the green wall: Why corporate programmes get stalled." *Corporate Environmental Strategy*, Vol. 2, No. 2, pp. 5-11.

Smithe, T. 1998. *The Green Marketing Myth: Tending Out Goats at the Edge of Apocalypse*. University of Toronto Press, Toronto.

Wong, V., W. Turner, P. Stoneman. 1996. "Marketing strategies and market prospects for environmentally-friendly consumer products." *British Journal of Management,* Vol. 7(3), pp. 263-81.

14

Psychology and the Natural Environment : Awakening for Averting an Impending Environmental Catastrophe

Govind Singh Kushwaha and *Vir Singh*

It is certain that if the growing environmental awareness is not translated into action for environmental conservation, we shall be in the clutches of an environmental catastrophe. Environmental awareness in recent times is one of the key achievements of humanity worldwide. However, this awareness does not seem yielding results, and humanity is not deriving full advantage from the natural system that could further strengthen this relationship and enhance symbiosis. The nature of this relationship is truly symbiotic. This relationship has been stronger in the past. But, in recent years this is somewhat disturbed. Human race has caused and appears to be bent upon causing severe damage to nature through accelerated pace of technological, industrial and agricultural development. Environmental damage in many instances is irreversible and also appears to be causing irreparable changes and negative trends. The damage to nature, in return, is creating devastating effects on human life and in development processes. The economic progress we are boasting of is unsustainable, because it is rootless. The human-environment relationship of our contemporary times is rooted into erroneous assumptions underlying the conceptualizations of man and the obvious scenario emerges in the form of the present-day environmental impoverishment and pollution, disturbed ecological balance and continued invasion on natural resources have assumed alarming magnitude.

Environment encompasses everything else—living as well as non-living. But environment includes several other aspects and, for human beings, it spells creative dimensions. *Environment*, broadly defined, includes not only physical spaces of all sorts, but socio-cultural contexts, neighbourhoods, institutions, and organizations—essentially any perspective in which people can find themselves. Sometimes, the term *proxemics* is used to connote the relationship between behaviour and the physical space in which it occurs (Hall, 1966; Watson, 1972). Stokols (1978) identified the central concern of *proxemics* as "the manner in which people use space as a means of regulating social interaction" and suggested that it subsumes work on privacy, personal space, territoriality, and crowding. Russel and Ward (1982) argued that interest in these topics has been heightened by concern about the ever-increasing world population, which threatens to make space an increasingly limited resource. Much attention has been given to the built environment—homes, offices, hospitals, schools—and to the effects of environmental stressors—extreme temperatures, humidity, oxygen deprivation, crowding, noise, air pollution, industrial mishaps—on human performance and well-being (Baum *et al.*,1982,1983; Baum and Singer, 1982; Cox *et al.*, 1982; Craik, 1973; Evans and Cohen, 1987; White *et al.*, 1982; Monat and Lazarus, 1977; Poulton, 1970; Welford, 1974).

Environmental Psychology has been recognized as a field of psychological research since the late 1960s (Craik, 1970, 1973; Heimstra and McFarling, 1974; Proshansky *et al.*, 1976; Stokols, 1978). E*nvironmental Psychology*, as Russell and Ward (1982) defined, is "that branch of psychology which is concerned with providing a systematic account of the relationship between person and environment". General or conventional psychology cannot be separated from environmental psychology except that the latter defines environment-human relationship and human behaviour *vis-à-vis* environment more conspicuously. As Stokols (1978) emphasizes, environmental psychology would seem to encompass all areas of psychology. Humans react to their immediate surroundings. Environmental Psychology attempts to respond to such interactions (Proshansky *et al.*, 1976).

Human dimension has profound impact on our systems, including natural systems. Nothing in this world can evade human impact.

Changes due to human interference can be localized and over time, they can assume global proportions. Human activities can change the whole environmental scenario, as we experience in the form of climate change in our contemporary times. Indiscriminate deforestation, alteration of land-use, rampant industrialization, intensive urbanization, creating a web of roads, etc., are examples whose primary impact is localized. But, cumulative effect takes the whole world into its grip. Global warming with serious consequences of climate change reveals the significant impact of human dimension at global scale.

Social and behavioural science attempts to explain the global-change phenomenon. Stern (1992) described the role of social and behavioural science in a global-change research agenda as improving understanding of how:

- human systems produce the proximate causes,
- changes in human systems might change the rate at which people alter the environment,
- people perceive changes in the global environment,
- people respond to the anticipation of global environmental change and are affected by experienced change, and
- changes in human systems might make people less susceptible to the effects of global environmental change.

Everyday new plans and programmes relating to natural resource management and environmental conservation are coming to the fore. These plans and programmes and their proponents make tall claims about considerable progress on the front of environmental conservation. However, we hardly experience a ray of hope in this regard. Bald truth is that environmental deterioration remains one of the most serious issues facing humanity posing serious challenge for the governing bodies to take concrete and decisive action. Economic considerations often become an eclipse of environmental considerations. Development woven with environment is often overstated, but the general experience is that sustainability stays on stake. Liberalization, Privatization and Globalization (LPG) processes have added fuel to fire. LPG processes are pushing the environmental issues on the periphery of our thinking.

The overall environment scenario hints at a future that is not sustainable, and a different response is required from the sciences

(Kates *et al.*, 2001). Not only is environmental sustainability an ecological crisis, it also includes the viability of socially shaped relationships between people and nature. The destruction of natural environments threatens the well-being of humans in ways we are only beginning to appreciate. We need sweeping changes in human behaviour to lessen our collective impact on the natural environment. We also need a better understanding of the human-nature experience and a more compelling language to express what we value and love. The magnitude and urgency of these problems present new challenges and opportunities for the social sciences in general, and for psychology in particular.

Human behaviour is largely responsible for environmental degradation. Nevertheless, such degradation sooner or later brings health hazards for human beings, and also hampers socio-economic progress. It is owing to this fact that psychologists have an important and vital role to play in analyzing the linkages between people and environmental problems and finding achievable and effective solutions (Oskamp, 2000a, b; Winter, 2000).

Global environmental changes, shortages, and threats which have direct bearing on the Indian natural environment include population growth causing stress on the environment, environmental pollution, rampant industrialization, annexation of forest lands for non-forestry activities, climate change and its consequences, consumption patterns, rapidly diminishing fossil fuel reserves, etc. Such are the grim issues India is faced with. Environmental conservation, ecosystem restoration and management, coping with water scarcity in many areas, reclamation of saline soils, improving water quality, controlling logging and mining, climate change mitigation, etc., are the major challenges India is facing.

Several workers and organizations (e.g., Union of Concerned Scientists, 1992, 1997; World Resources Institute, 2002; Millennium Ecosystem Assessment, 2003; Kennedy, 2006; World Wildlife Fund, 2006) have brought to the fore the facts associated with environmental threats, both emerging and pending. These environmental threats are very real and a cause of considerable scientific and public concern. They, in turn, reflect global environmental pressures and impacts. A whole set of these threats and inevitable linkages with psychological factors have been made available by noted psychologists.

Environmental problems, degradation, and global warming have been regarded as very pressing social issues and societal concerns, as well as environmental issues. They constitute important and indeed urgent areas of needed multidisciplinary collaborative work across the natural and social sciences.

Environmental Psychology may be of critical importance for improving the lot of global environment. It is of great strategic value. Diagnosis of environmental problems, policy decision-making, practical intervention, effective evaluation are the areas where psychological applications can apply in an effective manner (Schmuck and Vlek, 2003).

Gobster and Hull (2000), Syme *et al.* (2002), Vining and Ebreo (2002), Saunders (2003), Reser and Bentrupperbäumer (2005), and Dinsdale and Fenton (2006) spell out various roles environmental psychologists can play and are playing. Many psychologists, for example, work for or act as consultants to environmental management agencies, planning authorities, and government bodies, and advise on psychological and social considerations and issues. Psychologists can handle such things very well, for these would include the measurement and assessment of community attitudes, values and concerns, the relative effectiveness of differing communication and behaviour change strategies. The conventional "top-down" approaches of development often exclude environmental considerations. Psychologists can help switch over to a "bottom-up" or "the people side" approach of natural resource management or of development, which will ensure effective participation of local people in conceiving, designing and implementing projects of vital social, ecological, national, and global importance. Environmental Psychology, in essence, can help articulate principles of sustainable development and can lay the foundation of a sustainable future for the world.

Environmental Psychology underlines the essential human nature interconnectedness and it is potent to improve this relationship for the ultimate welfare of human beings. Applications of psychology are numerous. It is potent enough to avert the on-going processes of environmental catastrophe. Living in a secure and sustainable environment emanating from ecological amelioration is one of the most crucial urge embedded deep into human heart, mind and psyche. Psychology articulated into ecologically-oriented sustainable development directing us to usher in a sustainable future can make all the difference humanity has been aspiring for.

REFERENCES

Baum, A. and J.E. Singer (eds.). 1982. *Advances in Environmental Psychology:* Vol. 4. *Environment and Health.* Hillsdale, NJ: Lawrence Erlbaum Associates.

Baum, A.; R. Fleming, and L. Davidson. 1983. "Natural disaster and technological catastrophe." *Environment and Behaviour, 15,* 333-354.

Baum, A.; J.E. Singer and C.S. Baum 1982. "Stress and the environment." In G.W. Evans (Ed.), *Environmental Stress* (pp. 15-44). Cambridge, UK: Cambridge University Press.

Cox, V.C.; P.B. Paulus; G. McCain and M. Karlovac. 1982. "The relationship between crowding and health." In A. Baum & J.E. Singer (eds.), *Advances in Environmental Psychology:* Vol. 4. *Environment and Health* (pp. 271-294). Hillsdale, NJ: Lawrence Erlbaum Associates.

Craik, K.H. 1970. "Environmental Psychology." In K.H. Craik; B. Kleinmuntz; R.L. Rosnow; R. Rosenthal; J.R. Cheyne; & R.W. Walters (Eds.), *New Directions in Psychology* (pp. 1-121). New York: Holt, Rinehart & Winston.

Craik, K.H. 1973. "Environmental Psychology." *Annual Review of Psychology,* 24, 403-422.

Dinsdale, E.A. and M. Fenton. 2006. "Assessing coral reef condition: Eliciting community meanings." *Society and Natural Resources*, 19, 239-258.

Evans, G.W. and S. Cohen. 1987. "Environmental stress." In D. Stokols & I. Altman (Eds.), *Handbook of Environmental Psychology.* New York: Wiley.

Gobster, P.H. and B. Hull. (eds.). 2000. *Restoring Nature: Perspectives from the Social Sciences and Humanities*. Washington, DC: Island Press.

Heimstra, N.W. and L.H. McFarling. 1974. *Environmental Psychology.* Monterey, CA: Brooks Cole.

Kates, R.W., W.C. Clark; R. Corell; J.M. Hall; C.C. Jaeger; I. Lowe; J.J. McCarthy; H.J. Schellnhuber; B. Bolin; N.M. Dickson; S. Faucheux; G.C. Gallopin; A. Grubler; B. Huntley; J. Jager; N.S. Jodha; R.E. Kasperson; A. Mabogunje; P. Matson; H. Mooney; B. Moore III; T. O'Riordan; and U. Svedin. 2001. "Sustainability Science." *Science, 292,* 641-642.

Kennedy, D. (ed.). 2006. *Science Magazine's State of the Planet,* 2006-2007. Washington, D.C.: Island Press and American Association for the Advancement of Science.

Milleniunm Ecosystem Assessment. 2003. *Ecosystems and Human Well-being: Synthesis*. A Report of the Conceptual Framework Working Group of the Milleniunm Ecosystem Assessment. Washington, DC: Island Press.

Monat, A. and R.A. Lazarus (eds.). 1977. *Stress and Coping.* New York: Columbia University Press.

Poulton, E.C. 1970. *Environment and Human Efficiency.* Springfield, IL: Thomas.

Proshansky, H.M.; W.H. Ittelson and L.G. Rivlin (eds.). 1976. *Environmental Psychology: People and their Physical Settings* (2nd ed.). New York: Holt, Reinhart & Winston

Reser, J.P. and J.M. Bentrupperbäumer. 2005. The psychosocial impacts of visitation and use in World Heritage Areas: Researching and monitoring sustainable environments and encounters. In W.L. Filho (Ed) *Handbook of Sustainability Research* (pp. 235-263). New York: Peter Lang Scientific Publishers.

Russell, J.A., & L.M. Ward. 1982. "Environmental Psychology." *Annual Review of Psychology, 33,* 651-688.

Saunders, C.D. 2003. "The emerging field of conservation psychology." *Human Ecology Review*, 10, 137-149.

Stern, P.C. 1992. "Psychological dimensions of global environmental change." *Annual Review of Psychology, 43,* 269-302.

Stokols, D. 1978. "Environmental Psychology." *Annual Review of Psychology,* 29, 253-295.

Syme, G.J.; B.E. Nancarrow and B.S. Jorgensen. 2002. "The limits of environmental responsibility: A stormwater case study." *Environment and Behaviour*, 34, 836-847

Union of Concerned Scientists. 1992. *World Scientists' Warning to Humanity.*

Vining, J. and Ebreo, A. 2002. "Emerging theoretical and methodological perspectives on conservation behaviour." In R.B. Bechtel & A. Churchman (Eds.) *Handbook of Environmental Psychology*, pp. 541-558. New York: Wiley.

Welford, A.T. 1974. *Man Under Stress.* New York: Wiley.

Winter, D.D.N. 2000. "Some big ideas for some big problems." *American Psychologist*, 55, 516-522.

World Resources Institute. 2002. *A Guide to World Resources 2001-2002: People and Ecosystems: The Fraying Web of Life.* Washington, DC: World Resources Institute, United Nations Development Programme and World Bank.

World Wildlife Fund (WWF). 2006. *Living Planet Report 2006.* Gland: Switzerland: World Wildlife Fund, Zoological Society of London & Global Footprint Network.

Part III

CLIMATE OF CHANGE

Changing to Challenge Climate Change

15

Role of Van Panchayat Forests in Mitigating Carbon Emission in Uttarakhand

Vardan Singh Rawat, Ashish Tewari and *Y.S. Rawat*

Introduction

Over a time span of few years the heat balance of the earth was imbalanced. This happened due to sudden increases in Green House Gases concentration (GHGs) in the atmosphere. Increasing scientific evidence suggests that the impacts of warming will be more serious and will occur sooner than had previously been believed. Eleven of the last 12 years are among the warmest years recorded since 1850. The surface temperature has risen by about 0.74°C in the past 100 years and it could even rise up to by 5 °C by 2080 (IPCC, 2007).

Global climate change induced by the increased emission of green house gases is real, widespread and worldwide which is the predominant cause of the climate change. Earth's warming climate is estimated to contribute to more than 1,50,000 deaths and five million illnesses every year and this could be double in 2030. Change in climate affects earth ecosystems in number of such way such as; receding of Himalayan glaciers due to melting of ice; alteration in agriculture system; rainfall patterns; including worse droughts; water stress and flash floods; contaminations in soil and water bodies; settlement patterns; loss of biodiversity causing extinction of species, human health and so forth. Nearly 20-25 per cent of the annual atmospheric increase of about 8 billion tonnes of carbon is a consequence of deforestation, which results in the depletion of the carbon-sink.

Therefore, conservation of forests, including those under the control of local communities in developing countries, is an important component of overall climate strategy. However, the possibility of expanding carbon storage in forests has been identified as a potential measure to mitigate climate change by sequestering atmospheric carbon dioxide. Carbon sequestration is the long-term storage of carbon in ocean, soil, vegetation (especially forests) and geologic formations. Sink is used as a term to refer to the ability of forests, croplands, and grazing lands to sequester carbon. This process of using sinks is part of the international strategy to combat global warming by reducing atmospheric green house gases including CO_2. The important role played by forests in sequestering CO_2 from the atmosphere, and the livelihoods and environmental benefits that will be accruing to the local communities enable community forests to meet the dual objectives of Clean Development Mechanism (CDM) of sustainable development and emissions reduction.

Hence, there is growing interests in linking community managed forests to climate change. The community forests managed in accordance with Van Panchayat Act is a hybrid of state ownership and community responsibility. A major objective of Van Panchayat is to rejuvenate and manage patches of civil-soyam forests for local use. It also prevents neighbouring villages from intruding into this zone, once demarcated as a Van Panchayat forest.

Material and Methods

The selected sites are located in the Van Panchayat forests of Anriyakot and Bhatkholi villages in Lamgara developmental block of Almora districts at 29º 32′98″ to 29º 34′32″ N latitudes and 79º 41′ to 79º 43′2″ E longitudes. The elevation of these Van Panchayats varied from 1800 to 2000 m. The basic climate pattern is governed by the monsoon rhythm. The average annual rainfall varied from 32.93 mm to 51.59 mm. The mean maximum temperature varied from 17.31°C (December) to 27.87°C (June) in year 2009 and 17.34°C (December) to 28.90°C (June) in year 2010 while the mean minimum temperature varied from 2.18°C (January) to 14.87°C (June) in year 2009 and 5.00°C (January and February) to 8.50°C (October) in year 2010. Rock types mainly comprises of schist, micaceous quartizimeta morphism, plutonic bodies of granodiorites and granites (Valdiya, 1980). Altitudinally the study areas are located in temperate environment but latitudinally it comes within subtropical belt. The vegetation type mainly comprises Himalayan moist temperate oak forest and subtropical pine forest. The dominated tree species of both the Van Panchayats are *Quercus leucotrichophora, Pinus roxburghii, Rhododendron arboreum,* and *Myrica esculenta.* Information on socio-economic parameters was collected by questionnaires which were distributed to 30 per cent of the households in the Van Panchayat. The households were selected randomly on the basis of number of family members and categorized into small (<4), medium (5-9) and larger (>10). Estimations were also made by actual observations of the number of head loads removed daily, actual number of days in a year in which collection is done, number of hours and the frequency of collection by each family.

Four aspects on both the Van Panchayats were identified. At each aspect, trees were analyzed by placing randomly 10 × 100 m^2 circular quadrats. However, saplings, seedlings and shrubs were studied by using 10, 5 × 5 m^2 quadrats and herbs in 10. 50 × 50 cm^2 quadrats placed randomly. Tree layer biomass was estimated on the basis of allometric equations previously developed by Rawat and Singh (1988) and Chaturvedi and Singh (1987). Carbon stock and sequestration rate were estimated as 50 per cent of the dry weight of biomass and 50 per cent of net primary productivity, respectively (Hamberg, 2000; Brown, 2001).

Results

Some vegetational, socio-economic and soil parameters of the study area are given in Table 15.1. The total forest biomass in Anriyakot Van Panchayat along four aspects was 120.07 t/ha/yr in the first year which increased to 129.91 t/ha/yr, of which tree layer contributes 111.98±28.29 t/ha/yr in the first year and 119.74±29.0 t/ha/yr in the next year and the contribution of herb, shrub and litter was 8.09 t/ha/yr in the previous year and 10.17 t/ha/yr in the second year. Of the total biomass the above ground parts was 82.22±5.05 t/ha/yr in the previous year (2008) which increased to 88.55±5.11 t/ha/yr in the second year (2009), while the total below ground parts was 29.76±2.84 t/ha/yr in the previous year which increased to 31.19±2.96 t/ha/yr in the second year.

Of the total biomass the contribution of above ground part was 73.95 per cent and below ground part was 26.05 per cent. The

Table 15.1: Site Characteristics of the Anriyakot and Bhatkholi Van Panchayat Forests

Parameters	*Anriyakot Van Panchayat*	*Bhatkholi Van Panchayat*
Altitude (m)	1500-1800	1500-1700
Year of formation	1978	1976
Area (ha)	36.12	50
Dominant Vegetation	*Quercus leucotrichophora, Pinus roxburghii*	*Quercus leucotrichophora, Pinus roxburghii*
Total Tree density (in/ha)	150 to 490	193 to 324.3
Livestock population under each household	2-5	3-5
Daily requirements of fuelwood (kg/family)	5-7	5-7
Total household number residing around the Van Panchayat forest	12	15
Soil bulk density gcm^{-3} (0-100 cm)	1.09±0.007 to 1.36±0.03	0.82±0.06 to 1.36±0.005
Soil carbon% (0-100 cm)	1.00±0.29 to 2.73±0.51	1.41±0.54 to 2.97±0.46
Annual range of soil moisture (%) (0-100 cm)	7.53±1.06 to 29.72±1.15	6.52±1.09 to14.50±1.15

contribution of shrub, herb and litter layer was 3.90 per cent, 1.25 per cent and 2.68 per cent, respectively. While total forest biomass in Bhatkholi Van Panchayat was 50.1 t/ha/yr in the first year which increased to 58.94 t/ha/yr in the second year, of which the contribution of tree layer was 43.09±7.70 t/ha/yr in the first year and 49.90±8.50 t/ha/yr in the second year and the contribution of herb, shrub and litter was 7 t/ha/yr in the previous year and 9.03 t/ha/yr in the next year. Of the total biomass, the above ground parts was 31.9±1.37 t/ha/yr in the previous year (2008) which increased to 37.13±1.46 t/ha/yr in the next year (2009) while the total below ground parts was 11.2±1.04 t/ha/yr in the previous year which increased to 12.78±1.13 in the second year. Of the total biomass the contribution of above ground part was 74.39 per cent and below ground part was 25.61 per cent. The contribution of shrub, herb and litter layer was 5.72 per cent, 3.50 per cent and 6.11 per cent, respectively.

The average carbon sequestration rate in the present study in Anriyakot Van Panchayat forest was 3.90±0.52 t ha^{-1} yr^{-1}, while in the Bhatkholi Van Panchayat forest it was 3.41±0.69 t ha^{-1} yr^{-1}. The maximum sequestration rate was on south facing aspect (5.44 t ha^{-1}

Table 15.2: Total tree carbon stock and carbon sequestration rate in Anriyakot and Bhatkholi Van Panchayat

Aspect	*Carbon stock in 1st year (B_1) (t ha^{-1})*	*Carbon stock in 2nd year (B_2) (t ha^{-1})*	*Carbon Sequestration rate $\Delta B=B_2-B_1$ (t ha^{-1} yr^{-1})*
	Anriyakot Van Panchayat		
South-East	77.53	81.16	3.63
South	78.69	84.13	5.44
North	18.92	22.07	3.15
North-West	49.04	52.43	3.39
Mean	56.05	59.95	3.90
	Bhatkholi Van Panchayat		
South-West	26.37	31.67	5.30
North-West	29.54	32.82	3.28
East	17.76	19.79	2.03
North	12.78	15.79	3.01
Mean	21.61	25.02	3.41

Table 15.3 : Comparison of carbon stock and carbon sequestration rates across various forest types in the world with the current values

Forest type	*Carbon stock (t ha^{-1}).*	*Carbon sequestration (t ha^{-1})*	*Reference*
Tropical forest	285.0	2.30	Malhi (1998),
Temperate forest	125.0	4.19	Press *et. al.*
Boreal forest	67.0	1.40	(2000)
Predominant forest range of the world	50-150	—	Lal (2004)
All central Himalayan forests.	250-300	6.0-8.0	Singh and Singh (1985, 1992)
Seven Central Himalayan forests.	166.8-440.10	6.83-7.42	Rana *et al.* (1989)
Chir pine and banj oak non-degraded Forests.	173.7-262.6	5.06-6.66	Jina *et. al.*, (2008)
Anriyakot Van Panchayat forest (Mean carbon sequestration rate)	55.98-59.94	3.96	Present study
Bhatkholi Van Panchayat forest (Mean carbon sequestration rate).	21.61-25.03	3.42	
Dense mixed banj oak forest of Dhaili Van Panchayat.	255.7-260.2	4.5	Phhartiyal and Tewari (2008)
Young pure chir pine forest of Toli Van Panchayat.	69.5-74.0	4.5	

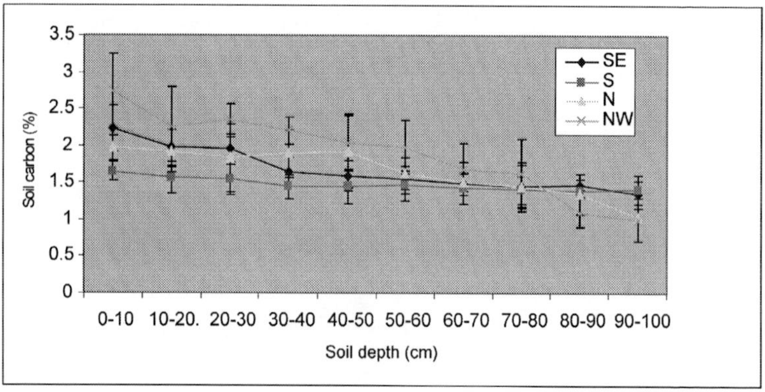

Fig. 15.1: Soil carbon per cent at Anriyakot Van Panchayat along different soil depth and aspects (SE-South East, S-South, N-North, NW-North West)

yr^{-1}) of the Anriyakot Van Panchayat forest followed by south-west facing aspect (5.30 t ha^{-1} yr^{-1}) of the Bhatkholi Van Panchayat forest. The minimum sequestration rate was on east facing aspect (2.03 t ha^{-1} yr^{-1}) followed by north facing aspect (3.01 t ha^{-1} yr^{-1}) of the Bhatkholi Van Panchayat forest. Data showed that in both the Van Panchayats, the highest amount of carbon was stored in the biomass of *Quercus leucotrichophora* trees (91.55 %) on the south facing aspect of the Anriyakot Van Panchayat. The soil organic carbon per cent varied from 1.00±0.29 to 2.73±0.51 in Anriyakot Van Panchayat (Fig. 15.1), while in Bhatkholi Van Panchayat this value varied from 1.41±0.54 to 2.97±0.46 (Fig. 15.2).

Discussion

Loss of forests is a significant contributing factor in climate change. However, the possibility of expanding carbon storage in forests has been identified as a potential measure to mitigate climate change by sequestering atmospheric carbon dioxide (FAO, 2001, DeFries *et al.*, 2000, Prentice *et al.*, 2000). The general goal of sequestration activities is to maintain ecosystem in the sink phase. Carbon stock and carbon sequestration rate varies among type and age of forest. A young forest would sequester more carbon than the old aged forest. A fast growing species would sequester more carbon than a slow growing species. This is not only true for newly planted forests but

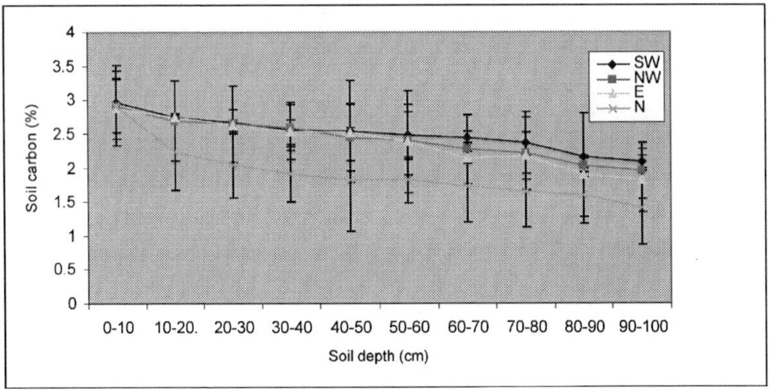

Fig. 15.2: Soil carbon per cent at Bhatkholi Van Panchayat along different soil depth and aspects (SW-South West, NW-North West, E-Eest, N-North)

Table 15.4 : Total tree, shrub, herb and litter biomass and carbon sequestration rate in Anriyakot Van Panchayat

Aspect	*Biomass 1st Year tha^{-1} B_1*	*Biomass 2nd Year tha^{-1} B_2*	*Change in biomass (t/ha) $B=B_2-B_1$*	*Carbon sequestration rate (t/ha/yr)*
Bole	42.30±7.70	45.28±7.82	2.98	1.49
Branch	22.92±6.68	25.64±6.97	2.72	1.36
Twig	11.61±4.12	12.09±3.88	0.48	0.24
Foliage	5.39±1.71	5.54±1.76	0.15	0.07
Total above ground	82.22±5.05	88.55±5.11	6.33	3.16
Stump root	24.22±7.02	25.15±7.42	0.93	0.47
Lateral root	4.86±1.30	5.05±1.40	0.19	0.09
Fine root	0.68±0.19	0.99±0.07	0.31	0.16
Total below ground	29.76±2.84	31.19±2.96	1.43	0.72
Total	111.98±28.29	119.74±29.01	7.76	3.88
Shrub	3.86±1.04	5.07±1.11	1.21	0.61
Herb	1.47±0.66	1.62±0.77	0.15	0.08
Litter	2.76±0.39	3.48±0.16	0.72	0.36
Grand total	120.07	129.91	9.84	4.92

also for established forests if their harvest rate and disturbance rate is less than the forest growth (Kagi and Schmidtke, 2005).

Table 15.5 : Total tree, shrub, herb and litter biomass and carbon sequestration rate in Bhatkholi Van Panchayat forest

Components	*Biomass Yr^1 tha^{-1} B_1*	*Biomass Yr^2 tha^{-1} B_2*	*Change in biomass (t/ha) $B = B_2 - B_1$*	*Carbon sequestration rate (t/ha/yr)*
Bole	18.57±1.45	20.54±1.47	1.97	0.98
Branch	8.00±2.40	10.03±2.47	2.03	1.02
Twig	3.51±1.23	4.52±1.39	1.01	0.51
Foliage	1.82±0.41	2.04±0.52	0.22	0.11
Total above ground	31.9±1.37	37.13±1.46	5.23	2.62
Stump root	9.12±2.68	10.08±2.88	0.96	0.48
Lateral root	1.84±0.38	2.30±0.42	0.46	0.23
Fine root	0.24±0.05	0.40±0.08	0.16	0.08
Total below ground	11.2±1.04	12.78±1.13	1.58	0.79
Total	43.10±7.70	49.91±8.50	6.81	3.41
Shrub	2.57±0.64	3.37±0.91	0.80	0.40
Herb	1.39±0.16	2.06±0.45	0.67	0.34
Litter	3.04±0.30	3.60±0.14	0.56	0.28
Grand total	50.1	58.94	8.84	4.42

Besides the above mentioned factors there are a number of other factors that would influence the carbon sequestration rate of forests such as temperature, rainfall, soil type and quality, microbial growth, predation, topographical features, anthropogenic disturbances and pollination, etc. Carbon sequestration can be achieved effectively through forest management and conservation practices. If one compares global warming to a fever of the planet than forest does not only function as a potential remedy but their destruction also contributes to further illness (Schone and Netto, 2005).

According to Upadhyay *et al.* (2005) revitalising degraded forest land and their soils in the global terrestrial ecosystem can sequester 50-70 per cent of the historic losses. Managed forests sequester more carbon than unmanaged forests nearing their climax stage as decay, burning, and die-back are balanced by the growth of plants (Upadhyay *et al.*, 2005). The present study illustrate that community forest management can be a viable strategy for reducing permanent emissions from deforestation, as the previous data reveal that the mean carbon sequestration rate for India (3.7 t ha^{-1} yr^{-1}) and Nepal (1.88 t ha^{-1} yr^{-1}), are close to 2.79 t ha^{-1} yr^{-1} or 10.23 CO_2 t ha^{-1} yr^{-1} under normal management conditions.

The average carbon sequestration rate in the present study in Anriyakot Van Panchayat forest in Almora district was 3.90±0.52 t ha^{-1} yr^{-1} while in the Bhatkholi Van Panchayat forest it was 3.41±0.69 t ha^{-1} yr^{-1}. In the present study the number of trees was more in young size girth class categories. Therefore, calculated carbon sequestration rate values are towards higher side. The Van Panchayats have great potential for future carbon sequestration if appropriate management practices are applied without human disturbance. The carbon sequestration rates of these Van Panchayat forests are within the ranges reported (2.4 and 5.6 t ha^{-1} yr^{-1}) earlier for forests of the Central Himalayan region by Press *et al.* (2000), Jina *et al.* (2008), Singh (2009), Raikwal (2008) (Table 15.3). The organic soil carbon per cent was higher at the top soil layer and declines as the soil depth increases, however, there was a little variation in the middle layers among all the aspects. On an average most of the soil texture of both the Van Panchayat was clay resulting in the higher value of soil organic carbon per cent.

Conclusion

In the Himalayan region, resources such as forests are better managed by local communities. Degraded forests have started rejuvenating through natural regeneration from stringent protection measures deployed by the local community. The present study shows that community managed forest generates both environmental and social benefits beside improving sustainable livelihoods and lifestyles, and encourage the people of the Himalaya to make a meaningful contribution to reducing global emissions and forest conservation. Environmental services include the conversion of forests from a source to a sink, improved watershed management, and biodiversity conservation. Social benefits include providing sources of livelihood for the rural population from community managed forest. If payment for carbon credit is made to these communities, added benefits may provide communities relying on the forests incentive to halt deforestation and opt for longer term benefits that are more sustainable in the long-run and those that may enhance their livelihood conditions. Carbon estimation is also necessary for monitoring the pool levels and in identifying strategies and management interventions to further improve efficiency.

Acknowledgements

Authors are thankful to G.B. Pant Institute of Himalayan Environment and Development (GBPIHED) Kosi Katarmal, Almora (Uttarakhand) for financial assistance to carry out the research and fieldwork. Authors are also thankful to community members of the Van Panchayat for their co-operation throughout the study period.

REFERENCES

Brown, S. 2001. "Measuring and monitoring carbon benefits for forest-based projects: Experience from pilot projects, can carbon sinks be operational?" *Resources for the Future (RFF)*. Washington DC, pp. 1-19.

Chaturvedi, O.P. and Singh, J.S. 1987. "The structure and function of pine forest in Central Himalaya. I. Dry matter dynamics." *Annals of Botany* 60: 237-252.

DeFries, R.; M. Hansen; J.R.G. Townshen; A.C. Janetos and T.R. Loveland. 2000. "A new global 1 km data set of per cent tree cover derived from remote sensing." *Global Change Biology*, 6, pp. 247-254.

FAO (Food and Agriculture Organization). 2001, Global forest resources assessment 2000: Main Report. *FAO Forestry Paper 140, FAO*, Rome, 479.

Hamburg, S.P. 2000. "Simple rules for measuring changes in ecosystem carbon in forestry-offset projects." *Mitigation and Adaptation Strategies for Global Change*, 5:25-37.

IPCC 2007. "Climate Change 2007: The Physical Science Basis Summary for Policy-makers." *Contribution of Working Group 1 (WG1) to the Fourth Assessment Report of the IPCC*. Pre-Publication edition. Geneva IPCC.

Jina, B.S.; P. Sah; M.D. Bhatt; Y.S. Rawat. 2008. "Estimating Carbon Sequestration rates and total carbon stockpile in degraded and non-degraded sites of oak and pine forest of Kumaun Central Himalaya." *Ecoprint*, 15: 75-81.

Kagi, W. and Schmidtke, 2005. "Who gets the money? What do forest owners in developed countries expect from Kyoto Protocol?" *Unasilva*, 222, 56: 35-38.

Lal, R. 2004. "Soil carbon sequestration impacts on global climate change and food security." *Science*, 304:1623-1626.

Malhi, Y.; A.D. Nobre and J. Grace. 1998. "Carbon dioxide transfer over a Central Amazonian rain forest." *Journal of Geophysical Research* 103: 593-631.

Phartiyal, P. and A. Tewari. 2008. "Challenges before Marginalized Hill Communities for Managing Community Forests." *Status of the Village Forest Council in Uttaranchal, India.*

Prentice, C.; M. Heimann; and S. Sitch. 2000. "The carbon balance of the terrestrial biosphere: Ecosystem models and atmospheric observations." *Ecological Applications*, 10(6), 1553-1573.

Press, M.C.; N.J. Huntley and S. Levin. 2000. "Ecology: Achievements and Challenge." *Blackwell Science*, Oxford, UK.

Rana, B.S.; S.P. Singh and R.P. Singh. 1989. "Carbon and energy dynamic of seven Central Himalayan forests." *Tropical Ecology* 30(2): 253-264.

Rawat, Y.S.; J.S. Singh. 1988. "Structure and function of forest in Central Himalaya. II. Dry matter dynamics, R. (1968). Ecology Workbook. *Annals of Botany*. 62: 413-427.

Schone, D. and M. Netto. 2005. "The Kyoto Protocol: What does it mean for forests and forestry." *Unasilva* 222, 56: 3-11.

Singh, J.S. and S.P. Singh, 1992. *Forest of Himalaya: Structure, Functioning and Impact of Man*. Gyanodaya Prakashan, Nainital, India.

Singh, S.P. and J.S. Singh. 1985. "Man and Environment: The Central Himalayan Case." *Biol. Mem.* 11(1): 47-59.

Upadhyay, T.P.; P.L. Sankhayan; B. Solberg. 2005. "A Review of Carbon Sequestration Dynamics in the Himalayan Region as a Function of Land-use Change and Forest/Soil Degradation with Special Reference to Nepal." In *Agriculture, Ecosystems and Environment*, 105: 449-465.

Valdiya, K.S. 1980. *Geology of Kumaun Lesser Himalaya*. Dehradun, Wadia Institute of Himalayan Geology, India.

16

An Effort towards Educating Hill Women about Indoor Air Pollution

Kiran Yadav, Neelam Bhardwaj and *Uma Melkania*

In order to awaken the people it is the women, who have to be awakened. Once she is on the move, the family moves, the village moves and the country moves forward.

—Jawahar Lal Nehru

Woman, God's most beautiful creation on earth, is a symbol of *W*—wisdom, *O*—origin, *M*—motherhood, *A*—affection, and *N*—nectar. The combination of all these qualities is the rising sun of India.

Women are not only the carriers of human race but civilization, sustainable development and culture rest on them. They are the best upholders of environment, ecological and social balance, yet a victim of man dominated patriarchal system, neglected in their day-to-day life, neither considered equal to man in wages nor in social status. Their triple role is silently appreciated but without economic recognition, regard and accountability.

Hill women have the dual burden of shouldering the responsibilities of looking after the family and of contributing to the family income. Their work pattern involves long hours of cooking, looking after the children on one hand and also working in the field and cattleshed, on the other (Srivastava, 1988).

Hill women who are the major contributors in raising the family are deprived of their rights due to insensitivity of the male dominated society towards many of their health problems. Hill women, due to lack of self-realization towards their own problems are silent sufferers. Thus, the conditions of this region can only improve if hill women are made aware about their own well-being and development. The fact is well supported by Wang and Dissanayake (1984) who defined development as a process of social change which has as its goal the improvement in the quality of life of all or majority of people without doing violence to the natural and cultural environment in which they exist and which seeks to involve the majority of the people as closely as possible in this enterprise making them masters of their own destiny. Communication plays a vital role in development process.

Although enough viable and adoptive technologies have been developed but many of these have not reached to the women residing in remote areas. The women in rural areas are deprived of minimum facilities of enlightenment and education. Since tradition plays a more important role in rural Indian scene, it is important to realize their differences in order to plan appropriate strategies for the spread of information among rural women. The formal system of education in the country covers only ten per cent of the total female population. What is worse is that the rate of wastage, stagnation and dropouts is much higher among girls than among boys.

Uttarakhand, a newly formed State, has thirteen districts out of which eleven are situated in hills and only two are in plain. So a vast majority of population belongs to hills. The hill women live in complete isolation from the outside world and they have limited contact with the neighbouring families as well.

Hill women work for longer working hours to attend to their household and field. Cooking and fuelwood collection are the major time taking jobs of the women. Hill women use traditional stoves or *chulhas* for cooking which emanate smoke. Since, kitchens in the most hilly homes are poorly ventilated, the smoke gets accumulated in the house. Prolonged exposure to high levels of pollutants has a very adverse impact on the health of women, who spend more than three hours for cooking everyday with biomass fuel. But the women are not aware of these health hazards caused due to indoor air pollution.

The awareness and knowledge amongst rural women can be generated through communication interventions. But it is the question of selective exposure of the audience to particular media messages. For generations, rural people are living in remote villages without access to modern means of communication. Women in hills and forest relied on the spoken words and traditional means of communication. This is because of their restricted mobility outside the home, lack of education and in some cases men's control over information and media. Until now, mass media have served largely as vehicles for top-down persuasion or as channel to convey information from authorities to the people. To rectify this situation low cost communication media can be used to increase the effectiveness of communication and to bring about greater participation of rural and urban poor in the development process (Wang and Dissnayake, 1984). Since the mass media is used for wider coverage but is impersonal in nature and may not be suited in the hill context.

A Case of Village Bhaluti

Bhaluti is one of the villages under Gram Sabha Jeolikote of Bhimtal block of district Nainital. Geographically, the village is located at the north-south border of district Nainital. At the entrance the village gives the impression as if it is a forest area as dense forest covers the whole village. These forests are the rich source of traditional medicinal plants and fuelwood. According to the village health worker, the nearby forest is rich in almost all types of medicinal plants used in *Ayurveda*. In addition, *Uteesh, Pine, Buransh, Hisalu, Kilmoda, Kafal, Bamboo, Ghingaru* are some of the commonly found trees in these forests. One special tree known as *Maalu* in local language was preferably used as fuelwood by the villagers. It was used because it burns for more than twelve hours and produces less smoke.

The settlement pattern of the village is unique in itself. One has to face the ups and downs of the hilly areas to reach the village. The households are scattered randomly and located quite a distance away from each other. This may be because of the availability of plain land to build houses. Farms of the villages are either near to the households or at a distance. The availability of water is also very poor. Out of 59 households only four were having their own tap connected through pipeline. Rest of the people either collect water from them or from nearby natural resource (water stream). There is one grocery shop in village Bhaluti. There is a common temple of 'Devi Maa', 'Shivji', and 'Hanumanji' in the village. The villagers of Bhaluti share common primary health centre, post office, Government High School and Government Inter College with the nearby villages. There is one Primary School and KVK in the village. There are two Self-Help

Groups running in the village namely 'Cheshta and Satyam'. But none of the women respondents from the selected village was the member of any of the self-help group.

Culture

The traditional beliefs and norms play an important role in various activities in the village. Hindu festivals like Holi, Deepawali, Shivratri, Janmashtmi are celebrated like in the other parts of northern India. Besides these *Ghugutiya* (Makar Sankranti), *Harela, Tulsi Ekadashi, Khatudua*, and *Ghew-tyaar* are some of the local festivals of Kumaun which are celebrated with great enthusiasm. During *Ghugutiya* ladies of each household prepare *ghugutee* (sweet dish prepared with wheat flour, jaggary and ghee) in the evening. The next morning children call the crow by saying "*kale kauaa aa ja ghugutee le ja*" (Oh! Black crow come and take away this sweet). Interestingly, hundreds of crows gather to eat *ghugutee*. The belief behind this activity is that through *ghugutee* you serve your forefathers.

Harela is celebrated in July-August. During *Harela*, five seeds including jowar, barley, wheat, rice and maize are sown in a basket filled with soil. Then on *Harela* day these seedlings are uprooted and then all the elder members of the family offer these seedlings to God before offering it to their family members and relatives. They give their blessings to them by saying "*Jeeray, Jagray, Bachray*" (live long and stay protected from all the evils). This is celebrated to welcome the new crops. During *Navratra* and one week before *Chaladi* (*Bari Holi/Dhulandi*) the women gather at one place and they sing and dance for one to two hours. In this way they celebrate the nine days of *Navratras* and *Chaladi*. The male folks celebrate *Holi* by organizing *Baithak* (collective gathering where they eat and drink local wine and dance late in the night). A day before *Chaladi* '*Chir*' is also celebrated. In this, youths of the village carry an umbrella like structure made up of bamboo. This structure is taken around the whole village. The women of each household worship and tie a strip of cloth around the wooden structure. Finally, it goes to the field where *holika dahan* takes place in the night. The next day villagers gather at a fixed time and at a common place and put *gulal* on each other's face. The young people put *gulal* on elder's face and touch their feet and seek their blessings. It was also noted that some boys and girls wrapped the cloth strip around their forehead and wrist after untying it from the wooden structure. Their belief was that this would protect them from evil power. From this belief one can assume that the villagers are superstitious. The villagers also believe in occultism, ghost, *tantra-mantra*, and *jhar-phoonk*. Many times *Jaagar* (a type of *pooja*) is also organized to thank, please and sometimes to wake up the gods. In this, the family or individual having some problem organizes *Jaagar*. People gather around a fireplace close to the house of the affected person. The *Harijan* play the *dhol* (a musical instrument). The music starts at a lower note and picks up slowly. Few selected people, not necessarily of the distressed family, sit around the fire. After some time, god comes in one of the person. This person act as god and blesses the gathering and tells the problem and its remedy to the distressed person. On the occasions of religious and cultural importance women decorate their houses by *aipan* (cultural design).

Dress and Language

The people of hill areas wear traditional dresses on special occasions. Men commonly wear *pajama* and *kurta* but nowadays pants, shirts are common in use, while women wear saree and blouse. During occasions like marriage, *namkaran sanskar*, *pooja*, *katha*, etc., women necessarily wear a special dress known as *pichhora* (a type of *dupatta*, which is in two colours with yellow background and red dots). Earlier, *haldi* (turmeric) was used for the yellow colour along with vermilion for red colour. Nowadays, readymade fancy *pichhoras* are very common. A big *nath* (nose ring) is also popular and common among women. *Nath* is considered as the symbol of their prosperity. While working, women tie their sarees in such a way that it is hardly exposed to dirt and does not hinder their work.

Status of Women

Women are in the centre of the family. Though the role of women is very important and crucial in hill economy yet they are still the invisible workers and lead a tough life. Women are actively engaged in all the activities right from the house to the fields. The day of women begins before the sunrise and continues even after the sunset. Women wake up early in the morning and go to the field or forest for fuelwood and fodder collection. Fuelwood collection is the major time taking job of the women of Bhaluti as they are using non-commercial energy sources for cooking. On average hill women spend more than 4-5 hours a day in fuelwood collection and its management. During fuelwood collection they cover a distance of about 4-5 kilometres everyday, carrying a load of 8-10 kg on their heads, backs, and necks.

Then they freshen up, clean the utensils and kitchen and start preparing meal for the family. They clean the house, take care of their animals by providing fodder, water and do milking and cleaning of the cattleshed. Outdoor work like grinding the wheat, buying the vegetables, oil and other household items are mostly performed by women. Some women prefer to do the fuelwood collection first and then go for other tasks.

The men in the hills usually do not work either in the fields or in the households. Very few cases were observed where a man was working. Unemployed youth, retired old men, daily labourers after their daily work, and even the school/college going youth usually gather near the shops where they drink tea, read the newspaper, gossip and play cards. Many of them drink local wine in the evening.

Interaction among women takes place during their work in the field and/or forest, kitchen, during washing clothes, etc. Sometimes, they make courtesy visit to their neighbours. There is no strict *parda* (veil) system in the village.

Role of Low Cost Communication Aids in Educating Women

Out of fifty-nine families in the village, all families were using non-commercial energy sources (like agricultural waste, animal dung, crop residues, etc.) for cooking. But they were ignorant about the health hazards caused due to indoor air pollution. On the basis of their need identification through certain Participatory Rural Appraisal (PRA) techniques like transect walk, resource map, and informal discussions were organized. Based on their information needs, low cost communication aids were designed.

Poster: The objective of poster was to create awareness among hill women regarding indoor air pollution. The message in the poster was written in local dialect, i.e., Kumauni.

Snake and Ladder Game: The snake and ladder game contained the information regarding health hazards, renewable energy sources,

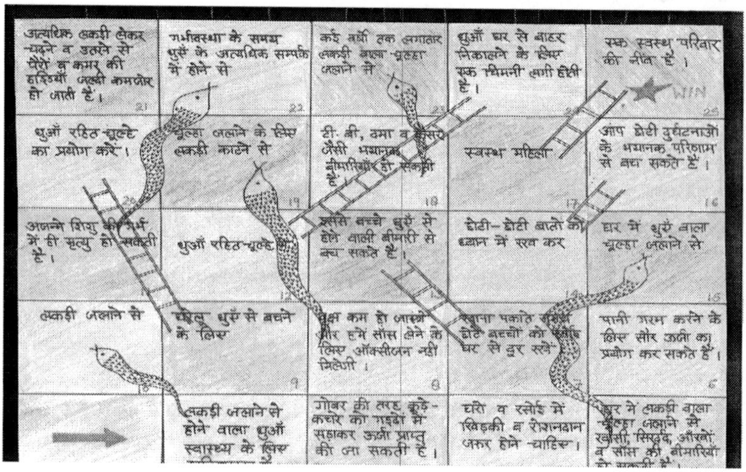

precautionary measures and other associated health hazards caused due to indoor air pollution. The game aimed at giving information in an informal way. The respondents themselves found it interesting and easy to play. The dimensions of the poster were 22×28 inches.

Flash Cards: The objective of flash cards was to provide an in-depth knowledge about renewable energy sources and health hazards.

The dimensions of flash cards were 11×14 inches. Total number of cards was eleven.

Drama: Drama is considered as one of the audio-visual aids. The objective of the drama was to create behavioural changes among hill women. Through drama, information regarding health hazards due to indoor air pollution was provided to the hill women. The drama was written in Hindi with local terminologies.

Major Findings

The data for the present study was collected from January to March, 2008. The salient findings of the study were:

- Need assessment was done through informal discussion with elderly women and women ward members of the Gram Sabha. Certain PRA techniques like non-participant observation, transect walk, resource mapping and daily routine diagram were used. It was concluded that fuelwood collection was predominantly done by women folk of the village. Wood collection was generally done in isolation this might be because of their differential daily routine. Women walk as much as four to six kilometres of hilly areas with a heavy load of six to

Cover page of flash cards

eight kilograms on their heads, backs and necks. They also carried their babies during wood collection.

- Maximum number of respondents (47.06%) had low level of knowledge regarding resources of indoor air pollution; majority of the respondents (50.98%) had low level of knowledge regarding health hazards caused due to indoor air pollution. In case of knowledge regarding renewable energy sources majority of respondents (50.98%) had low knowledge level.
- Findings related to the overall knowledge of the respondents revealed that majority of the respondents (45.09%) had low level of knowledge.
- Results of effect of Low Cost Communication Aids in terms of gain in knowledge showed that maximum number of respondents (58.82%) had medium level of knowledge regarding resources of IAP followed by high level of knowledge which was increased from 9.80 in pre-test to 25.49 per cent in post-test.
- Most of the respondents (84.31%) belonged to the category of medium knowledge level regarding health hazards caused due to IAP, maximum number of respondents (60.78%) were having medium level of knowledge regarding renewable energy sources.
- Findings related to the total knowledge gain of the respondents revealed that majority of the respondents (74.51%) had medium level of knowledge gained. Maximum knowledge gain was observed in the category of resources of IAP with mean gain score of 9.67.
- Education was positively and significantly correlated with gain in knowledge. Among the communication characteristics formal social participation was negatively correlated with gain in knowledge. Informal social participation and localiteness-cosmopoliteness were positively and highly significant with gain in knowledge.
- Feedback sheet was filled by the respondents at the end. Poster was found to be most appealing and left long lasting effect because the message was written in their local dialect, i.e., Kumaoni. Information given through flash cards was found to be very important and interesting. The game of snakes and ladder was also preferred by them because it gives

the information in an easy and compatible manner. The respondents found drama to be interesting and much closed to their real life situation. They express their views about Low Cost Communication Aids. They also suggested some things to make the programme and aids much more effective. They told that the aids were interesting, related to their real life situations, attractive, contain important issues and were easy to understand.

The results of the study are supported by Singh (1993) which emphasized that drama could bring about behavioural changes among the respondents. According to Pareek (1960), it is a method of studying the nature of certain role by acting out its concrete details in a conceived situation that permits of better and more objective observation. Drama captures people's imagination like no other television format and the half hour episodes used the power of drama to impact health education message. The technique is known as 'edutainment'. It conveys educational messages through entertainment programme (FAO, 1999).

Conclusion

The tradition of communication is as old as the civilization. Although electronic media has invaded every walk of life, yet in traditional societies interpersonal communication commands the supreme place. In our villages, majority of the population is still illiterate and poor. Therefore, interpersonal communication will continue to dominate and it is difficult to provide any other substitute. The principle of "seeing is believing" is the base that signifies the importance of visual communication. Villagers can be stirred to action with visible proof. Ignorance, illiteracy and poverty for centuries eroded the self-confidence among villagers. They believe that they are poor because God wants them to be poor. They hesitate to admit that they can also improve their socio-economic status by adopting various methods of communication. They believe in miracles although they have rarely seen one. Thus, when the villagers actually see the role of low cost communication aids in increasing the quality of life, they have to realize the power of these communication means.

Therefore, these aids have special significance in convincing the hill women about the superiority of these low cost communication aids. This is a small effort but the problem is not so small. So, to rectify this situation we need several such efforts which can bring these destitute, downtrodden and deprived women into the mainstream of development with honour, dignity and better health.

REFERENCES

FAO. 1999. "Voice for Change, Rural Women and Communication" prepared by Silvia Balit.

Pareek, U. and Y.P. Singh. 1960. "Socio-economic Characteristics of Key Communicators." *Sociological Bulletin*, XV, 1.

Singh, Ranjit. 1993. *Communication Technology for Rural Development*. B.R. Publishers and Distributors (P) Ltd. Delhi-110007.

Srivastava, S. 1988. "Women of Rural India: An Overview." *Yojana*: 4, pp. 4-7.

Wang, Georgette and W. Dissanayake. 1984. "Culture, Development and Change: Some Explorative Observations." In G. Wang and W. Dissanayake (eds.). *Continuity and Change Communication Systems*. New Jersey: Ablex.

17

Towards Nutritious India : March towards Slow Foods

Rita Patnaik and *Gita Singh*

> *Slow food unites the pleasure of food with responsibility, sustainability and harmony with nature.* —Carlo Petrini

> *Slow food made me realize how important it is for farmers, producers, cooks and consumers to work together in order to defend our agricultural heritage.*
> —Madieng Seck, Agronomist, Senegal

India is amidst a paradox with malnutrition with millions suffering from actual hunger as well as hidden hunger arising from lack of micronutrients such as vitamins and minerals at one end of the developmental spectrum. On the other hand, we have the largest epidemic of obese and diabetic people. The fast food culture which is sweeping across the world remains no longer an exception in India. With urbanisation and women empowerment building in pace, a completely new era has come into the picture where the term nutrition is at stake, where nutritious food has ultimately taken the crude form of fast foods or junk food! Invariably it is a burger, pizza, fries, cola drinks and the lot. Each of these adds to the calories and fat intake devoiding the diet of the essential vitamins and minerals.

In this era of modernisation we have forgotten our age old traditional foods of our grandmother which, apart from providing nutrition, also gave immunity to the body. So there is a need for a march forward from fast foods towards slow foods. Loads of applauses goes to Dr. Vandana Shiva of *Navdanya*, a non-profit organisation, who has spearheaded the slow food movement in India and revived the concept of slow foods which is in fact the tradition of our country. Her noble idea to link producers and consumers and to promote "good food through good methods" is worth praising.

Slow Food Movement

More than twenty years ago, when Italian Carlo Petrini learned that McDonald's wanted to erect its golden arches next to the Spanish Steps in Rome, he developed an impassioned response: he helped found the Slow Food movement. Since then, Slow Food has become a worldwide phenomenon. Petrini says, "We aren't eating food. Food is eating us".

Large-scale industrial agriculture has run rampant and penetrated every corner of the world. The price of food is fixed by the rules of the market, which have neither concern for quality nor respect for producers. People have been forced into standardized, unnatural diets, and aggressive, chemical based agriculture is ravaging ecosystems from the Great Plains to the Kalahari. Food has been stripped of its meaning, reduced to a mere commodity, and its mass production is contributing to injustice all-over the world. In *Terra Madre*, Petrini shows a solution in the thousands of newly formed local alliances between food producers and food consumers. And he proposes expanding these alliances—connecting regional food communities around the world to promote good, clean, and fair food. The end goal is a world in which communities are entitled to food sovereignty—allowed to choose not only what they want to grow and eat, but also how they produce and distribute it.
(http://www.slowfood.com/international/food-for-thought/publications/79786/terra-madre-forging-a-new-global-network-of-sustainable-food-communities/q=123).

Slow food, thus, is the antithesis of fast food. While fast food represents all that is fancy, takes less time to cook but is full of empty calories devoid of nutrients especially vitamins and minerals; slow foods are in fact our age old traditional foods representing our culture and diversity. Slow foods represent the whole foods with plenty of fibre, vitamins and minerals, releasing energy slowly into the system. These are in fact the protective foods that chronically reduce blood sugar levels and improve insulin sensitivity and insulin activity. When these protective foods have subsequently vanished from our diets, it has resulted in the so-called diabetic epidemic along with other diseases across the globe including India.

March forward towards slow foods is indeed a step forward to see what is in our environment right at our doorsteps nutritionally sound, healthy, that too at affordable prices. Our age old concept of cereal-pulse (rice-dal) combination seems to become extinct with soaring prices of pulses. In many households proteins from pulses are now being replaced with carbohydrates from potatoes. And yet there are some forgotten and forbidden foods, namely the millets grown in different parts of the country, starting from sea level in south to up in the mountains in the North. These are finger millet, barnyard millet, foxtail millet, and little millet which have high nutritional value and are one of the cheapest sources of dietary energy, in the form of proteins and carbohydrates along with high quantity of essential vitamins and minerals. Due to the high proportion of fibre, millets also have hypoglycemic effects which make the products of these crops suitable for consumption by diabetic and heart patients. High calcium and iron contents in finger millet make it suitable for making baby food products.

Why is it that the so-called vitaminaceous and nutritious green leafy vegetables which are available at the doorstep being abandoned not only by the poor but also the elite? Why is the fast food culture becoming deeply entrenched into our society? Why are overweight, obesity, diabetes and hypertension becoming the curse of the nation? Why cannot we inculcate good habit of taking fruits everyday? Why are we forcing our children into this world of fast foods ending into the epidemics?

Navdanya's Organic Foods

Navdanya's Organic Food movement has also captured roots in the world, as well as in the minds of people the world over. Dr. Vandana Shiva, a renowned environmentalist, ecofaminist, organic farmer and a revolutionary, is sparking wonderful ideas about food. "I see food as a gift to us from all the sentient beings who share the life with us", says Vandana Shiva. "Food is not just a commodity for sale", Vandana reiterates. "And when we treat food as a commodity we can see what the final consequences are if we look at the industrial agriculture of the United States."

Navdanya's Organic Food movement is capturing the attention of masses everywhere and is also influencing agricultural policies in a big way. In support of the organic farming vital for promoting the processes of climate change mitigation, Vandana Shiva says, "Organic farming in a way fulfils everyone of the principles of Gross National Happiness. It reflects cultural values. It promotes the protection of the natural environment. And in the area of agriculture, the key elements of the natural environment are biodiversity, diversity of the species, the soil, the water and the air. And then elements become more important with climate change, because the climate is changing". (http://www.sji.bt/assets/PDFs/Speech-of-Dr-Vandana-Shiva.pdf).

Celebrating Diversity of Foods

"Diversity is prosperity" is a traditional adage. *Baranaaja* culture raised by mountain farmers is a unique testimony of it. *Baranaaja* literally means twelve foodgrains intercropped with finger millet as the base crop. Amaranth, buckwheat, kidney beans, horse gram, black soybean, black gram, green gram, cowpea, rice bean, adzuki bean, sorghum, and *cleome* are the main crops intermixed with finger millet. *Baranaaja* provides a unique example of how a mountain farmer cultivates agrobiodiversity. This management helps provide the maximum number of food items ensuring balanced diet from the minimum area of land. If the degree of agrobiodiversity were directly proportional to the level of mountain farmers' food security, *Baranaaja* culture would hold the key. Cultivating diversity of foods such as in the *Baranaaja* system of the mountains is celebrating cultivating slow foods.

The main thrust of conventional interventions in the mountains such as in Uttaranchal is to replace the *Baranaaja* culture with monocultures of white-seeded soybeans. Soybeans as a cash crop are being projected as a panacea for the land-based economy of the mountains. This crop was introduced recently as one of the packages of the Green Revolution. Soybean, in fact, has never been an ingredient of local diets, nor has it fetched more money for the farm families. Farmers in Uttaranchal, who switched over to soybean cultivation in some areas, generally barter their produce for salt or rice. Transformation of the *Baranaaja* crops into soybean monoculture has also created livestock problems as the introduced crop, unlike all major mountain crops, does not provide fodder for livestock (Singh, 2005).

Baranaaja crops occupy only the rainfed and otherwise the least fertile areas of cropland where other moisture-demanding crops would fail to grow. The local farmers, using complementarity and synergistic associations amongst different food-yielding species, have evolved agrobiodiversity-centred agricultural practices; the wonderful *Baranaaja* culture being its glaring example. Such agrobiodiversity management enhances sustainability of the agroecosystem and raises the food security level of the farming community (Singh, 2005).

The only answer to all these is the option to switch our gears and take a U-turn towards slow foods, our traditional foods, our whole foods rich in fibre and nutrients. Yes, of course still patronizing the ideals of Baba Ramdev inculcating the *yoga* culture and imbibing vegetarianism let us all take an oath to lead a nutritious slow food for a healthy, creative and happy life.

REFERENCES

http://www.sji.bt/assets/PDFs/Speech-of-Dr-Vandana-Shiva.pdf

http://www.slowfood.com/

http://www.slowfood.com/international/food-for-thought/publications/79786/terra-madre-forging-a-new-global-network-of-sustainable-food-communities/q=123

Singh, V. 2005. *Agrobiodiversity, Sustainability and Food Security in the Himalayan Mountains: An Uttaranchal Perspective*. Gorakhpur: GAG.

18

Agri-based Community and Environmental Resistant Movement : A Case of the Bagur-Navile Tunnel Project, Hassan District, Karnataka

D. Jayadeve Gowda

Introduction

Bagur-Navile tunnel project is located at Bagur Hobli of Channarayapatna taluk in between Bagur and Navile, which covers the farmers of over 40 villages of Bagur Hobli. The project was aimed at the improvement of water supply and for the irrigation purpose to the people of Tumkur and Mandya districts. Unfortunately, the tunnel which passes through Bagur-Navile, due to seepage of groundwater into tunnel, bore wells and tanks in the area have gone dry. The idea of constructing a tunnel was mooted during the tenure of the Late Devraj Urs—former Chief Minister of Karnataka, with so many opposed to the support of political leaders of Hassan district. The tunnel was constructed below 200 feet has led to the depletion of ground water table.

The tunnel, 9.76 km long and 4.5 km wide, runs between Bagur-Navile tunnel from A.G. Ramachandra Rao Channel of the Hemavathi Project. The project went on 15 years and by 1991 the adverse effects became apparent. Hence, fertile land of many villages has become wasteland. The yield of 8,000 coconuts per year per acre recorded earlier has now gone down to only about 2,000 coconuts, i.e., a staggering reduction of 4 times in the production rate. Similarly, decline in the productivity of other agricultural crops has also been observed. Many families depending on the fishing are in problem due to depletion of water in lakes. With this background, an investigation was carried out to:

- find out the real causes of the Bagur-Navile Tunnel Project problem;
- look into problems of the farmers;
- bring to the fore the obstructions caused by political leaders and various organizations;
- find out remedies to overcome the problem; and
- find out the effect on the social conditions of local people.

Significance of the Study

Hassan district has created so many prominent political leaders including Chief Minister and Prime Minister of India. People of the district are mainly depending on agriculture and animal husbandry for their livelihood. The most of the western part of the district is mainly affected by Bagur-Navile tunnel project. It has disturbed farmers ecologically, politically and economically. There is a big conflict between development and conservation of natural habitat. But with the power of politicians, the voice of environmentalists and farmers was not heard by the implementing authorities. The implementing authorities are very keen to implement the project.

With the construction of second largest tunnel in Asia, the Bagur-Navile tunnel for about 9.7 km, have created a big hazard on land fertility, production of agricultural and horticultural products. That has progressively decreased the watertable by seepage of water into tunnel. Crop production has been reduced 6 times and plantation was wilted drastically. Crop loss of about Rupees 30 crore was assessed by irrigation and horticulture departments.

This research engages with the problems of 40 villages and their participation in maintaining the good health of soil and their water holding capacity of the study area. This research is about the understanding how cropland is being affected by the influence and negligence of the political leaders of Hassan district and how to come out with a common solution for the welfare of Bagur-Navile project sufferers.

Movement against the Project

There was a big protest movement started by the people of 40 villages with the support of communist leaders and local leaders in 1996-97. But this movement was ruthlessly suppressed by the district administration and the political leaders. People participating in the movement were also tortured by police force without giving any consideration to them. With this background, the partial relief has been made by the Karnataka Government. The sufferers of the affected area are not satisfied with this relief which has been made by administrative machineries of the government. They have expected total development of the affected villages. Therefore, they are interested in continuing their agitation in future. Here is a particular political party and its leaders gave a big support to the district administration for suppressing the agitation for their personal benefit.

Conclusion

India needs to take environmental concern seriously and follow the laws, although laws cannot be applied retrospectively and goalposts cannot be constantly shifted. It is also a great idea to institutionalize a local stake in the future growth of the region. But these causes are not going to be served by converting the Environment Ministry into an instrument of blackmail and recrimination. Bagur-Navile tunnel project cannot be accepted on ecological, environmental, socio-cultural and economic bases. How can the ecological damage caused by the project can be healed, how can economic loss done so far and to be accrued to the farmers in future be restored, and how should the culprits—politicians, engineers, contractors, administrators, etc. behind the project, and police responsible for farmers' tortures—be brought to justice are the real issues emanating from this project.

Acknowledgement

I am thankful to UGC for granting funds on MRP and also thank Dr. A. Thimmegowda, Principal of A.V.K. College for Women, Hassan, Karnataka.

19

Role of Community Radio in Climate Change

Arpita Sharma

Introduction

The effects of climate change can be already perceived in some fields, the most often in rural areas as natural disasters, drought and flooding, decrease of biodiversity, invasive species, loss of agricultural productivity, and even food shortage. Climate changes included global warming. Global warming is projected to have significant impacts on conditions affecting agriculture, including temperature, carbon dioxide, glacial run-off, precipitation and the interaction of these elements. The overall effect of climate change on agriculture will depend on the balance of these effects. Assessment of the effects of global climate changes on agriculture might help to properly anticipate and adapt farming to maximize agricultural production. The agreement is expressed in the 1996 report of the Intergovernmental Panel on Climate Change (IPCC), an international body of leading natural and social scientists sponsored by the United Nations Environment Programme and the World Meteorological Organization. According to the panel's report, an increase in atmospheric concentrations of greenhouse gases equivalent to a doubling of carbon dioxide (CO_2) will force a rise in global average surface temperature of 1.0 to 3.5 degrees Celsius by 2100. Average precipitation also will rise as much 10 to 15 per cent because a warmer atmosphere holds more water. Agriculture is extremely vulnerable to climate change. Higher temperatures eventually reduce yields of desirable crops while encouraging weed and pest proliferation. Changes in precipitation patterns increase the likelihood of short-run crop failures and long-run production declines. Although there will be gains in some crops in some regions of the world, the overall impacts of climate change on agriculture are expected to be negative, threatening global food security. The impact of climate change is not good among the rural communities. Majority of Indians are dependent on farming as a main occupation. Climate change has bad impact on the agriculture as well as on health. It is necessary to communicate the message related to environment protection among the rural poor through the proper medium. Various research studies have confirmed that Community Radio is an effective and cheapest medium for the communication of the message.

Climate Change

Climate change is a change in the statistical distribution of weather over periods of time that range from decades to millions of years. It can be a change in the average weather or a change in the distribution of weather events around an average. Climate change may be limited to a specific region, or may occur across the whole Earth. Rural areas have a significantly higher carbon production rate per person than urban areas. This is principally due to generally higher income and per capita consumption and because of the higher emissions from heating homes and transport in rural areas. About 63 per cent of the developing world's total population lives in rural areas, according to World Bank estimates and they include nearly 75 per cent of the approximately 1.2 billion people who are trapped in extreme poverty. The expected rise in global temperatures will affect human health, comfort, lifestyles, food production, economic activity, and residential and migration patterns. As global temperature rises, atmospheric circulation patterns are likely to change with alterations in the frequency and seasonality of precipitation and an overall increase in the rate of evaporation and precipitation. Coupled with the associated general rise in temperature, such changes in the water cycle will affect water availability, agricultural activity, flood protection practices, infrastructure planning and natural habitats. If the intensity of the hydrologic cycle increases, some scientists believe that humanity and natural systems may experience much more severe weather related events such as droughts, floods and extremely severe tropical cyclones. One estimate is that an effective CO_2 doubling may increase the intensity of tropical cyclones or hurricanes as much as 40 per cent.

Effects of Climate Change

Climate change and agriculture are interrelated processes, both of which take place on a global scale. An even greater sea of troubles, though, is the one encroaching specifically on agriculture. Worldwide, farming is a significant source of greenhouse gas emissions, accounting for a sizable share of the estimated 20 per cent of total emissions that result from land-use. Agriculture's vulnerability to climate change and the this implies grave consequences for the world's poor and hungry. The climatic change could affect agriculture in several ways: Climate change affects the productivity, in terms of quantity and quality of crops, agricultural practices, through changes of water use (irrigation) and agricultural inputs such as herbicides, insecticides and fertilizers, environmental effects, in particular in relation to frequency and intensity of soil drainage (leading to nitrogen leaching), soil erosion, reduction of crop diversity, rural space, through the loss and gain of cultivated lands, land speculation, land renunciation, and hydraulic amenities, adaptation, organisms may become more or less competitive, as well as humans may develop urgency to develop more competitive organisms, such as flood resistant or salt resistant varieties of rice. Agriculture has been shown to produce significant effects on climate change, primarily through the production and release of greenhouse gases such as carbon dioxide, methane, and nitrous oxide, but also by altering the Earth's land cover, which can change its ability to absorb or reflect heat and light, thus contributing to radiative forcing. Land-use change such as deforestation and desertification, together with use of fossil fuels, are the major anthropogenic sources of carbon dioxide; agriculture itself is the major contributor to increasing methane and nitrous oxide concentrations in earth's atmosphere.

Community Radio

Community radio is often the only medium available in the interiors. The content is in the hands of community and people participate in the programming addressing issues that matter to them. We chose 15 villages in Himachal Pradesh and gave the locals complete freedom to air their views while we acted as facilitators. Initially, there was a lukewarm response but now we are flooded with people's requests.

Community radio is an effective medium to disseminate the information among the rural area. It provides the information regarding climate change, agriculture, health, sanitation and hygiene. A community radio station can be roughly defined as a short range radio station that caters to the information needs of communities living in surrounding areas. CR stations often involve local community members in programme production, centred around topics including discussions on civic amenities in the area, health and hygiene, advice on common economic activities such as agriculture, and even local folk songs and cultural events. This notion of community radio is somewhat restrictive because it specifically defines communities along geographic boundaries. A community can, however, be geographically dispersed and Internet radio can be used to connect community members not collocated with each other. But we will go with this simple definition for now. Although technically a CR station is just like a normal AM or FM radio station, its focus on locally relevant content, often socially motivated, distinguishes CR from private and commercial radio stations most of which generally broadcast songs and entertainment related content. Community radio, although new to India, has been known to be an empowering medium since a long time. Radio Sagarmatha in Nepal played a seminal role in restoring democracy. Dialogues on CR between the Hutu and Tutsi tribes in Rwanda helped reconciliation efforts after the genocide. All across the world, in Brazil, Bolivia, Poland, Thailand, community radio has provided a voice to the people to air their thoughts and opinion. India has been a late comer to community radio. Although educational institutions were allowed to set up a CR station on their campus, typically referred to as a campus radio station, a specific community or group or non-profit organization was not given a broadcast license. However, in late 2006, a new CR policy was drafted and non-profits were allowed to apply for licences as well. India has seen a steady growth in CR stations since then, and as of October 2009 there are now almost 45 CR stations out of which some 8 stations are being run by non-profit organizations.

Role of Community Radio in Climate Change

Climate change is a change in the world's climate; it is any long-term significant change in the "average weather" that a given region experiences. Average weather may include temperature, precipitation and wind patterns. (*Wikipedia*). The term climate change is used interchangeably with "global warming" and "greenhouse effect" but is a more descriptive term. It refers to the build-up of manmade gases in the atmosphere that trap heat, causing changes in weather patterns on a global scale. Community radio has the capacity to offer valuable help to the Nation during disaster and epidemics in creating public awareness through local dialect to save lives by providing specific warning. Community radio can provide information in detail from local level to the administration and other community and help lifeline agencies in taking due preparedness measures. Moreover, to minimize mass sufferings from different forms of natural and manmade disasters, community radio can take instant initiatives to disseminate warnings through local dialects for better understanding.

Providing Information on Climate Change : The Role of Community Radio in Africa

In the book entitled *Fighting Poverty—Utilizing Media in a Digital Age*, Grace Githaiga said, "Community radios have facilitated on improved livelihoods through information in different important areas such as trading and business development, breaking myths on HIV/AIDS, improving health, agriculture etc." It is expected, therefore, that the same medium would be used to disseminate information on climate change in the language of the people and bring it down to their level. Indeed community radio can serve as effective channel through which farmers, economic operators and other citizens can access information on global warming. As a means of information and communication, community radio in Africa has a great role to play in information dissemination on climate change. It has the duty to promote the rights of the people to communicate, assist the free flow of information and opinions on climate change. The Community radio can focus discussions on environmental degradation and harmful local practices like bush burning and deforestation, which worsen global warming. The CR should provide the right access to information and serve as interactive medium where the citizens can ask questions and raise their concerns about climate change. The Community radio should serve as a medium to speed up the process of informing the people about climate change thereby serving as a catalyst for change. The Community radio should in partnership with national, regional and international research institutions design programmes that will provide the required information on weather variability such as change in rainfall pattern and increase in temperature as well as offer suggestions to farmers and the citizens on climate change, what actions to take to safeguard the environment and how to cope with the situation. Since agricultural extension workers are primary sources of information for farmers, the Community radio should liaise with them to reach farmers with relevant information on droughts, flooding, the intensity or severity of storms actions they can take to ensure good farming practices for improved harvest. When all these have been accomplished, the CR would have played its expected role in taking information on climate change to the people.

Conclusion

Community radio is a type of radio service that caters to the need and interest of certain area, broadcasting content that is popular to a local audience but which may often be overlooked by commercial or mass media broadcast. Community radio is also referred as rural radio, cooperative radio, participatory radio, free radio, alternative radio, popular radio, educational radio, community FM, association radio and bush radio in different parts of the world. The basic characteristic of Community Radio is that it is owned by the community, to serve the needs of the people. Most significantly, it is participatory in nature. People actively take part in formulating the station's policy, strategy and programme content. Community radio aims to change social conditions and improve the quality of cultural life through meaningful and relevant programmes. More than half of the percentage of the population is based on farming as a main occupation. This is the main reason that in the rural areas farmers are affected more through the climate change. The Earth's climate system constantly adjusts so as to maintain a balance between the energy that reaches it from the sun and the energy that goes from Earth back to space. Community Radio is an effective medium to disseminate the information regarding the climate change, health and various other aspects.

REFERENCES

amarcwiki.amarc.org/upload/documents/Winet_Nigeria.doc
en.wikipedia.org/wiki/Climate_change
en.wikipedia.org/wiki/Climate_change_and_agriculture
envfor.nic.in/cc/what.htm
www.climateandfarming.org/
www.climatescience.gov/Library/sap/sap4.../Synthesis_SAP_4.3.pdf
www.ifpri.org/.../climate-change-impact-agriculture-and-costs-adaptation
www.ifpri.org/pressroom/briefing/impact-climate-change-agriculture
www.prb.org/Articles/2007/ClimateChangeinRuralAreas.aspx.
www.sap43.ucar.edu/documents/Front_Matter.pdf

Index